T0185828

Astronomers' Universe

Series Editor
Martin Beech, Campion College, The University of Regina,
Regina, SK, Canada

The Astronomers' Universe series attracts scientifically curious readers with a passion for astronomy and its related fields. In this series, you will venture beyond the basics to gain a deeper understanding of the cosmos—all from the comfort of your chair.

Our books cover any and all topics related to the scientific study of the Universe and our place in it, exploring discoveries and theories in areas ranging from cosmology and astrophysics to planetary science and astrobiology.

This series bridges the gap between very basic popular science books and higher-level textbooks, providing rigorous, yet digestible forays for the intrepid lay reader. It goes beyond a beginner's level, introducing you to more complex concepts that will expand your knowledge of the cosmos. The books are written in a didactic and descriptive style, including basic mathematics where necessary.

More information about this series at http://www.springer.com/series/6960

Carlos Martins

The Universe Today

Our Current Understanding and How It Was Achieved

 Springer

Carlos Martins
Centro de Astrofísica da Universidade do Porto
Porto, Portugal

ISSN 1614-659X   ISSN 2197-6651   (electronic)
Astronomers' Universe
ISBN 978-3-030-49631-9   ISBN 978-3-030-49632-6   (eBook)
https://doi.org/10.1007/978-3-030-49632-6

This Springer imprint is published by the registered company Springer Nature Switzerland AG.
The registered company address is: Gewerbestrasse 11, 6330 Cham, Switzerland

Preface

> You do not receive an education any more than you receive a meal. You seek it, order or prepare it, and assimilate and digest it for yourself.
>
> Frank Rhodes (1926–2020)

Over the past 10 years, I have had the pleasure of giving various different short courses (between 10 and 25 contact hours each, depending on the case) to three types of audiences. This book grew out of the contents of these lectures, with the encouragement of many people who attended them.

The first group are high school teachers, in the context of refresher courses for current teachers regularly organised by U. Porto and by ESERO's[1] partner in Portugal, Ciência Viva, and mainly taught at CAUP. Periodically taking some such courses is mandatory for high school teachers to progress in their careers, and naturally, the courses are broadly aligned with the goals of the national physics and chemistry school curriculum. (Unfortunately, Portugal has no national astronomy curriculum!) These are formal courses, including a final assessment (typically a written exam), and are validated by a national pedagogical committee before being offered. About 70% of the teachers who take them are physics and chemistry teachers, while the others are mathematics and biology/geology teachers.

The second group are U. Porto students (about two-thirds of them studying science or engineering and the rest from all other areas), taking complementary

---

[1]https://www.esero.pt/.

courses offered by the CAUP Training Unit. Most of these are undergraduate students, with occasional M.Sc. or Ph.D. students. The students can take these courses at any point of their studies, on a voluntary basis: a final written exam is not compulsory, although it can be credited if the students choose to take the exam and pass (typically about one-third of the students take the exam, and almost all those who take it pass).

The third group are bright high school students (in the last 3 years before university, roughly 15–18 year olds), mostly in the context of an astrophysics summer school which I created in 2012 and have been organising every year since then at the Corno de Bico Protected Landscape Area[2] (in the Paredes de Coura municipality, in the Northwest of Portugal), the AstroCamp.[3] Initially, this was only aimed at Portuguese students but has now grown into an international school, accepting applications from 42 eligible countries and supported by international partners like ESO in addition to several national partners.

These courses are given to relatively small groups (in Portuguese and to a maximum of 40 students in the first two cases or in English and to a maximum of 20 in the third case), allowing for detailed interaction with the students and for instant feedback from them. They are mostly given using slides, supplemented by the blackboard whenever needed. Occasionally, I have also used various parts of this material as stand-alone popular talks, mostly when visiting high schools, for example, during World Space Week (in this case, the groups are larger, typically 50–120 students). About half of this material is on modern astrophysics (physics of stars, relativity, the standard Big Bang model, physics beyond the standard model, etc.), while the other half is on the history of astronomy (and, to a lesser extent, that of physics).

After many queries from students on where they could find additional information on the various topics discussed, the time has come to further organise some of these materials into a book.

My aim is to give the reader an overview of our current view of the Universe, of how we gradually developed it, and of how outcomes of current research— both my own and by others—might still change this view. For this, one must bring together concepts in physics and astronomy, including some of the history of both of them. The historical part may seem of lesser interest to the average reader, but it is important to understand where we are and how things may develop in the future. In fact, I strongly believe that anyone taking

---

[2]http://www.cornodebico.pt/portal/.
[3]http://www.astro.up.pt/astrocamp/.

physics or astronomy today and considering a future professional career in this field should be exposed to the history (and philosophy) of these subjects—and naturally the same applies to any other scientific subject. It is the responsibility of universities to provide such possibilities in the degrees they offer.

The level of the book is not technical. Although I do not particularly like the word, it could be called 'descriptive', in the sense that the goal is always to highlight the crucial physical concepts. I often emphasise to students that if one understands the key ideas in a given context, one can always work out the maths later on (if and when it is needed), while trying to start with the maths without understanding the physics is far harder. Therefore, there will be very little explicit maths, and most of it will be concentrated into a few sections. That said, the level of the book will not be as simple as that of typical popular science books: the aim is to present things rigorously, and although I will simplify many things I did my best to avoid oversimplifying.

My guess is that the 'average' readers should be undergraduate students (not necessarily studying astronomy or physics, but still with a broad interest in these areas and/or taking an introductory course on it). The only prerequisite is a qualitative knowledge of basic physics concepts, at the high school physics level. The book (or some chapters thereof) could therefore be useful for various undergraduate introductory astronomy and physics classes. This level should also make the book relevant for high school teachers who may need to teach some parts of this material or simply acquire some background knowledge with which to answer questions from the more interested and curious of their students. (At least in the Portuguese school system, some astronomy topics are actually taught by biology/geology teachers, who sometimes have a limited knowledge of astronomy or physics.) For the same reason, the book should be appealing to bright high school students wanting to learn more physics and astronomy.

The book aims to put together materials from several different short courses but broadly includes one part on the history of astronomy (and to a lesser extent of physics) and another one on modern astrophysics and cosmology. The early chapters are not purely 'historical', as they could have been written by a historian of astronomy or physics. For the record, I am a working scientist, not a historian, and there are many aspects of the history of science which historians find fascinating that are of very little interest to me—naturally, the opposite is also very probably true. The point of delving into the history of the subject is that of introducing, in a historical context, concepts that are important for our view of the Universe today. By discussing how they emerged and subsequently evolved, one can more easily understand their modern relevance.

As a convenient way to summarise some of the concepts discussed in the book, I provide a short multiple-choice quiz, together with some suggested points to think about. The reader can think of the latter as invitations to write short essays on these topics, as a means to further consolidate knowledge. The answers to the multiple-choice quiz can be found online. Some of these questions have been used previously in the exams I set for students taking the course. Finally, some suggestions for further reading are also provided in an overall bibliography.

Porto, Portugal   Carlos Martins
March 2020

Acknowledgements

As mentioned in the preface, this book grew out of material from various courses I have given in the last decade or so. I am grateful to the many students in these courses, whose questions and comments gradually allowed me to develop the material in relevant and interesting directions and pitch it at an appropriate level for the various intended audiences.

I am grateful to Paulo Maurício, with whom I co-taught many editions of the History of the Universe course, for many interesting discussions about the historical aspects, and to my students, especially Ana Catarina Leite and José Ricardo Correia, for discussions on the more modern aspects.

Many thanks also to Francesco Chiti, Liliana Sousa, Marlene Körner, Paloma Thevenet, Pedro Amaral, and Siri Berge, all former AstroCamp students, who have read several chapters and provided useful comments and suggestions.

Last but not least, I am grateful to Angela Lahee and the Springer team for their patience during the completion of this project and the many helpful suggestions during the various stages of its development.

This work was partially financed by FEDER—Fundo Europeu de Desenvolvimento Regional funds through the COMPETE 2020—Operacional Programme for Competitiveness and Internationalisation (POCI), with grant number POCI-01-0145-FEDER-028987, and by Portuguese funds through

FCT—Fundação para a Ciência e a Tecnologia in the framework of the project PTDC/FIS-AST/28987/2017 (CosmoESPRESSO).

Cosmology and Fundamental Physics with ESPRESSO

Contents

Acronyms

BCE	Before the Common Era
CAUP	Center for Astrophysics of the University of Porto
DNA	Deoxyribonucleic acid
EEP	Einstein equivalence principle
ELT	Extremely Large Telescope
EPR	Einstein–Podolsky–Rosen
ESA	European Space Agency
ESERO	European Space Education Resource Office
ESO	European Southern Observatory
GPS	Global Positioning System
GUT	Grand Unified Theory
HEO	High Earth orbit
HST	Hubble Space Telescope
ISS	International Space Station
LEO	Low Earth orbit
LHC	Large Hadron Collider
LIGO	Laser Interferometer Gravitational-Wave Observatory
MACHO	Massive compact halo object
MEO	Medium Earth orbit
NASA	National Aeronautics and Space Administration
NEO	Near Earth object
NOAA	National Oceanic and Atmospheric Administration
RAS	Royal Astronomical Society
VLT	Very Large Telescope
WIMP	Weakly interacting massive particle

1

Introduction

We will start with a brief discussion of the scientific method, highlighting what distinguishes science from other human endeavours and reflecting on how what we now call science originated historically. We then take a more detailed look at these origins in the specific case of astronomy, in Egypt and Babylon, also mentioning the development of tools such as calendars and clocks. Finally, we reflect on the importance of scientific literacy (and its mathematical sibling, numeracy) in the modern world, and briefly mention astrology in this context.

1.1   The Scientific Method

> Science is a way of trying not to fool yourself. [...] The first principle is that you must not fool yourself, and you are the easiest person to fool.
> Richard Feynman (1918–1988)

One of the more noteworthy and alarming paradoxes of our modern civilisation is that the more our everyday lives rely on science and technology, the less the common person knows about them. If you're not sure about what I mean, I suggest the following exercise. Try to spend 24 h of your life using only the technologies you understand—by 'understand' I mean that you can explain how they work, in a simplified but otherwise accurate way, to a teenager.

© Springer Nature Switzerland AG 2020
C. Martins, *The Universe Today*, Astronomers' Universe,
https://doi.org/10.1007/978-3-030-49632-6_1

For example, if you don't understand how your mobile phone works, and how it captures your voice and image and sends them to the nearest tower, and from there ultimately to the person you're talking to, then don't use a mobile phone. If you don't understand how the engine of your car (or bus, or train) works, then don't use them during that day. If you don't understand how an airplane flies, then don't fly. And we could go on to computers, digital cameras, microwave ovens, and so on.

Notice that I'm not talking about building these tools—they are all tools to enhance our own capabilities, one way or the other. I'm simply talking about understanding, at a non-technical level, what basic principles (of physics, chemistry, biology, maths) enable them to do for you what they are supposed to do. So whenever you reach out to grab one of these tools, pause for a second and ask yourself whether you understand how it works. I'm sure that you will have a very interesting 24 h.

This highlights the importance of science in technology in your daily life. And since it is clearly affecting your life, it should also make you think about what this thing is that we call science. Certainly, you will have read or heard many news stories saying that scientists have just discovered or announced something new. That could be something that is easy to grasp (possibly about a new species of mammal in the Amazon forest), but it could also be something completely outside your everyday experience, for which you have no previous experience, even if it makes you curious—maybe something about the big bang or black holes?

So how do scientists do what they do, and arrive at their conclusions on such topics? And, more importantly, why should you know about these things, and believe the things that scientists say? Clearly, you're not going to go to the Amazon rainforest to check that the newly discovered mammal is really there, or check some mathematical calculations, or a lab experiment, or a computer simulation, so either you simply trust the claim, or you have to hope that someone else will check and confirm (or refute) such claims for you.

Answering the second question is easy enough. Most of the challenges and threats that we will all be facing in the coming decades, and which are hotly debated today have a science component at their core (as a partial list, consider global warming, terrorism, pandemics, and 'alternative medicines'). Clearly, they are not purely scientific problems—they involve economic, political, ethical, and moral aspects, among others—but they do involve science at their core. And knowing the science is crucial for anyone to participate in the corresponding debate. If you don't know the underlying science, you may simply ignore or exclude yourself from the debate, and you will be easily manipulated, either by people who do know more than you and want you to shift your opinion in a particular direction for their own purposes, or even by people that know less than you (although they may think that they know more).

Most cultures articulate their world picture, and systematically transmit it to the next generation, through mythology or religion. In the beginning of the seventeenth century, the Western world decided to do something different, by articulating our view of the world through science, and particularly through astronomy and physics. Nevertheless, science has always been confined to a small number of specialists, and despite four centuries of development of our education systems most people receive a comparatively poor exposure to science, either as children or later on as adults.

In this sense, our society hasn't been doing particularly well, and indeed the development of seemingly key ingredients has been rather slow. To take a simple example, the term 'scientist' was only coined fairly recently: it was first used in print by William Whewell in 1834, in his review of Mary Somerville's book 'On the Connexion of the Physical Sciences'. (As an aside, quite a few terms we now use in physics and chemistry—such as ion, electrode, cathode, or anode—were suggested by Whewell to Michael Faraday.) Also, science as a profession, in the sense that a restricted number of people are extensively trained to be able to carry out this activity (and eventually be paid for doing it, as well as training the next generation) is only a late nineteenth century development.

How did science emerge? We will look into the particular case of astronomy shortly, but for the moment let's note that, from a historical perspective, societies before ours had two main motivations that eventually led to what we now call science. The first one is abstract and conceptual, and in its proper historical context it is essentially theological—to behold God's plan for the Universe. The second one is practical. If you are a farmer and your survival in the coming months depends on producing enough food to get through the next winter you will be highly motivated to understand the cycle of the seasons so that you can sow and harvest at the appropriate times. You thus need to develop an accurate calendar, leading to astronomy. Similarly, the need for reliable maps, methods of navigation, and all sorts of tools which enhance the natural capabilities of the human body are starting points for various other branches of physical science.

As a specific example, think for a second about the night sky. For early societies this provided a calendar for farmers and a map for sailors, but also a home for the gods and a repository for countless mythological stories for everyone to tell and remember. Indeed, the night sky was like a primitive society's television—or, in more modern parlance, YouTube. What else could

Fig. 1.1 False colour map showing the intensity of skyglow from artificial light sources. Credit: P. Cinzano, F. Falchi (University of Padova), C. D. Elvidge (NOAA National Geophysical Data Center, Boulder). Copyright Royal Astronomical Society. Reproduced from the Monthly Notices of the RAS by permission of Blackwell Science

you look at during those long nights? On the other hand, in our modern societies very few people can see a truly dark sky, free from light pollution (this is in fact impossible in Europe, see Fig. 1.1), and even comparatively simple things like seeing the Milky Way or the Andromeda galaxy with the naked eye can be a challenge. Very few people have had the opportunity to see a sky as dark as the ones Galileo saw just four centuries ago. This is particularly unfortunate since the night sky is the only part of our environment (and cultural heritage) that is common to everyone—all human beings, no matter when or where they live(d) see basically the same night sky.

It is interesting that most societies prior to our own can be neatly classified into one of the two sides of this conceptual/practical divide. The canonical examples are Classical Greece and the Roman Empire. In the former the prevailing view was that the world should be understood but not changed, and therefore any knowledge which had practical applications was deemed inferior to purely abstract knowledge. On the other hand the Romans were an eminently practical civilisation and adopted whatever worked, without ever worrying (or even thinking) about why it worked.

What is peculiar about our society is that we have somehow optimally combined the two motivations in the early seventeenth century and this led to the development of modern science. On the one hand, several developments in the previous century (by Copernicus, Tycho, and many others, which we will describe in the coming chapters) required the development of new theoretical paradigms in astronomy and physics. On the other hand, new practical tools such as the telescope and the microscope appeared at precisely this time, allowing the testing and further development of these paradigms.

How, then, does one do science? The starting point in the scientific method is a belief in the objective validity of science. This includes three different aspects that one must accept (or, perhaps more accurately, assume). Firstly, that Nature really does exist outside of and independently of us. To put it simply, you should accept that the Universe existed before you were born, and will continue to do so after you die. Secondly, that there is some set of laws of Nature which are objectively valid, without regard for our preferences or expectations. And thirdly, that we can progressively discover and understand these laws. Note that the third is conceptually different from the second: it could be the case that such laws exist but are entirely beyond our reach. These are the assumptions that every scientist is—at least implicitly—making. Science itself cannot prove the correctness of these assumptions, but as one proceeds one can gather supporting evidence for them. Historically, one could say that these assumptions were first made in a systematic way in Classical Greece around 2500 years ago, and more clearly reasserted in early seventeenth century Europe.

The scientific method is an iterative way of generating consistent knowledge about how the Universe works, gradually identifying old ideas which prove inadequate and replacing them by new ones, on the basis of observations of and experiments in the real world. One starts by observing a particular aspect of the Universe that is of interest, and formulates a starting hypothesis which is consistent with these observations. This hypothesis is then used to make further predictions, which are in turn tested by further experiments or observations. The process is then iterated until there are no noticeable discrepancies between the hypothesis (and the underlying theory, if it already exists) and the experiments. Once this is successfully achieved, the hypothesis is validated and accepted as a new theory, or added to an existing one. By this process our knowledge about the physical world gradually grows and we acquire a deeper and more accurate understanding of the aforementioned set of laws of Nature.

An interesting question is what is the ultimate source of scientific knowledge. Two different answers are provided by rationalists (who emphasise reason and intuition) and empiricists (who emphasise experience and observation). Examples from the two camps are René Descartes (1596–1650) and David Hume (1711–1776), respectively, but this division is prevalent throughout the history of science itself, for example in Plato versus Aristotle or in Newton versus Galileo. Einstein is interesting in this regard, because he developed his special and general theories of relativity in opposite ways.

There are several different but inter-related concepts here that are worth distinguishing. A law is a scientific hypothesis for which there is an ample collection of experimental and/or observational evidence. A theory is an underlying conceptual framework which is able to explain a set of experimental and observational results and the corresponding laws, and which additionally predicts the results of new observations and experiments that can be done subsequently. A theory with a limited range of applicability, which is clearly perceived to be a first approximation, or which still lacks extensive testing, is sometimes called a model.

A first important aspect of science is its iterative (one could say trial and error) nature. One thing that the history of science teaches us (and which is crucial for prospective scientists to be aware of) is that in many circumstances scientific progress is the direct result of the realisation that we were asking the wrong question. When this happens one can sidestep the original question by tackling a different but related one, and not uncommonly one also finds that the first question was actually irrelevant. We will see examples of this later in the book. In some sense one could say that what distinguishes science from other human endeavours is not what it allows us to know and how we deal with that knowledge, but how we confront what we still don't know.

A second important aspect of science is an asymmetry between the confirmation and the refutation of a theory or hypothesis. Refutation is always a possibility and when it happens is logically certain, but there is no logically valid way of proving (in the mathematical sense of the word) the truth of a theory from the agreement of its predictions with any finite number of observations or experiments. In other words, no hypothesis can be proved absolutely true (that would require testing it in all possible ways under all possible circumstances), but there is always the possibility that it can be proved false, if one of its predictions is not verified by a particular experiment.

We can never be certain that a theory is correct, only that refuted theories are incorrect. So effectively, what one has are degrees of confidence in the validity of each theory, which increase with every new observation or experiment consistent with them and drop to zero if one experiment refutes them. For example, being at mid-latitudes in the northern hemisphere I am extremely confident that the Sun will rise towards the east tomorrow morning (and indeed I can easily calculate—or look up online—the time and direction, in Porto for example, for the event), but I cannot prove that this will happen. Maybe the Sun will be destroyed overnight by Vogons working on a hyperspace bypass, in which case it won't rise tomorrow morning. (If you don't know who the Vogons are and why they are building hyperspace bypasses, you are not reading enough.)

It follows from the previous paragraphs that to be scientifically useful, hypotheses must be falsifiable. Any and all scientific theories are, by their very nature, in constant danger of being proved wrong by new data or observations. This is a crucial positive aspect of scientific research: it ultimately provides the means for constantly improving our knowledge about the Universe. Therefore scientific truths are always qualified, and never absolute. New discoveries that change our view of the Universe can occur at any point, either by falsifying previously held theories or hypotheses, or by setting limits on their domains of applicability, and thus highlighting the need for more encompassing ones.

To put it in a different way, the distinguishing feature of a scientific theory is not that it can be verified but rather that it can be falsified: it must be able to make further predictions, beyond the observations that led to its development, that can be subjected to further testing. Only vulnerable hypotheses and theories that include this element of risk, in the sense that their predictions must be specific enough to be incompatible with some possible results of the further tests, can be counted as supporting evidence for the theory.

From this we also see that a further important aspect of the scientific method is measurability: the theory's predictions must be specific enough to be quantitatively measured, even if at a particular point in time the available technology is insufficient to reach the accuracy necessary to measure the predicted effects. And another important aspect is reproducibility: the predictions made by a theory should apply to all the phenomena and circumstances it claims to describe (one can't have a theory that works on Mondays but not on Saturdays).

Reading the above you may reflect on how it correlates with the public perception of what science is, and in particular on how scientists are portrayed in everyday life, in particular by the media. It is worthy of note that no scientist (or, at least, no scientist worthy of that name) has or even claims to have the monopoly over truth. However, what scientists can do better than anyone else is to find out when hypotheses or theories are wrong and simply can't work. Indeed this is what scientists are specifically trained for, and most scientists, throughout most of their careers, spend their time excluding hypotheses and showing that many of them can't work. It is only relatively rarely that a scientist will discover something genuinely new about the way the Universe works.

That said, there are two caveats to bear in mind. The first caveat is that one must not reverse the burden of proof: anyone making new claims must also supply evidence in support of those claims. If you believe that Santa Claus does exist, the rest of the scientific community is not obliged to explicitly test and refute your claim: rather, it is your task to show what evidence you have supporting that claim. And the second caveat is that the fact that we do not know how something might work does not prevent us from finding out whether or not it works. For an example of this, have you ever had surgery which required you to be given general anesthesia? If you have, you may reflect on the fact that the number of people on the planet who understand how general anesthesia works, at the basic biochemical and neurophysiological level, is exactly zero. Local anesthesia is very simple to understand, but general anesthesia involves the brain and central nervous system, and understanding how it works at a fundamental level is a far more complex task. And yet, nobody doubts that it does work, and it has been routinely used across the world for decades, with a vast set of empirical data on the appropriate dose for the circumstances of each patient.

Finally, it is worth remembering that scientists are also human beings, and have their own preferences and biases which affect the way science is done. If a scientist has been working with a previously successful model or theory, it is often the case that data alone will not be sufficient to force him or her to completely abandon it and start again from scratch. Instead, such theories are often modified or reinterpreted in the light of the new developments. An example of this is the relation between Newtonian physics and General Relativity, to which we will come later in the book.

Max Planck famously said that a new scientific theory does not replace an older one by convincing the supporters of the latter that the new one is conceptually better, or simpler, or more accurate, but rather because the older generation that had been trained in the old one eventually dies and the new generation that replaces it is now familiar with and accepts the new theory. (Or, to put it more succinctly, science progresses with every funeral.) When they are faced with a choice between competing alternative theories, scientists often appeal (implicitly or explicitly) to further selection criteria, in addition to the results of experiments or observations. Examples of these include beauty, symmetry, or economy of explanation. Naturally these are, to a large extent, aesthetic concepts, which cannot be quantified and even lack a generally agreed definition. But these factors do enter scientific debates.

Could science disappear? In other words, given our modern reliance on science and technology and our choice of science as the lingua franca to articulate our view of the world, is this necessarily a continuous and irreversible process, or could it be stopped and reversed by the forces of irrationality? The simple answer is that history demonstrates that this has happened several times in the past, in other societies, from Classical Greece to Medieval China. And in our modern society science is threatened in many ways.

A scientifically and technologically advanced society comes without any guarantee that irrational thought will disappear. A particularly worrying trend is the fact that 'fringe phenomena' are widely spread by the media, indeed, more so than science itself. Think of astrology, creationism and intelligent design, global warming denial, anti-vaccination activists, and a whole slew of the so-called alternative medicines. It is clear that our society has a deep problem of scientific illiteracy, which is exploited by those wanting to spread scientific misinformation for ideological reasons.

The scientific community must do its part to address this problem. There are two components to it: one is making science accessible to the general public, and the other is explaining the process by which science is done. And academic institutions must provide greater incentive and facilities for their researchers to do so. Those working with younger students, and starting the process of training the next generation of citizens (and of future scientists), have a crucial role to play, too. As long as our society continues to report reality through science, and to rely on it to improve our daily lives, there's a society-wide obligation to make science accessible to everyone. What is at stake is not just individual sanity and critical thinking skills, but ultimately social cohesion.

1.2 The Dawn of Astronomy

The first indications of a desire to understand the world around us are provided by the mythology of each society. Indeed, these myths typically include an account of how gods or other beings created the world. Such myths can therefore be said to be the deep roots of modern astronomy and cosmology. But the myths also have practical consequences, and shape the way in which each society organises itself.

Nevertheless, having a set of anthropomorphic deities capable of interfering in human affairs has a huge drawback: it necessarily leads to a capricious world, in which one cannot hope to make reliable predictions about future events, because divine intervention can occur unpredictably at any point.

Thus one key step in the development of science is to overcome the innate tendency of interpreting natural phenomena as personified and divinised. Strictly speaking, such a step was, as far as we know, only decisively taken in Classical Greece. We will discuss this in the next chapter. Here we will go further back in time, to discuss the origins of observational astronomy.

Sedentarization and the ensuing development of agriculture provided a key incentive to make careful observations of the Sun, the Moon, and also the planets and stars in the night sky, for example to track the passing of the seasons in order to determine the best times to sow and harvest. And inevitably, such observations would lead to the discovery of the regularity of some astronomical phenomena. In fact, this is even attested by prehistoric monuments such as the Almendres Cromlech, Stonehenge (see Fig. 1.2), Newgrange, and many others. Although those who built them left no written records, it is clear that the building of such monuments, which obviously required a substantial multi-generation construction effort, is witness to an existing belief in the regularity of certain astronomical phenomena, such as the solstices.

The fact that astronomy arose in Middle-Eastern civilisations is not an historical accident. The various civilisations that flourished in this part of the world differed in several ways (some of which will be discussed in what follows), but they shared at least four common factors which made them ideally placed for the development of astronomy.

The first and most obvious one is that this was (and still is) an area of the planet where the sky is often clear, so making frequent observations is not particularly difficult. Secondly, sedentarization and the concomitant specialization of different members of the society led to elites who had enough free time (either ex-officio or simply by their own personal choice) to undertake this systematic study of the heavens. Thirdly, they had written languages, which enabled them to record and conserve their observations over very long periods (as opposed to having to rely on the fallible memory of individuals) and thus gradually accumulate an extensive set of data. Fourth, they had considerable mathematical knowledge and enough tools (today we would call them algorithms) to allow them to look into the accumulated data, notice important patterns or regularities, and thereby make practical use of them to try to predict future motions of the celestial objects.

Fig. 1.2 The Almendres Cromlech (Évora, Portugal) and Stonehenge (Wiltshire, England) (Public domain images)

In what follows we will discuss how different local factors shaped the development of astronomy (and the corresponding early cosmogonies) in two adjacent but contrasting regions, Egypt and Mesopotamia. From here this knowledge gradually spread to Greece and thence to the whole Mediterranean, which one can arguably consider to be the first civilisation.

1.2.1 Ancient Egypt

In Egypt, the regular patterns of the heavenly bodies are seemingly reproduced by the regular cycle of the Nile, with its annual flooding from July to October. Indeed, Herodotus called Egypt 'the gift of the Nile'. In this sense life in Egypt didn't visibly change, and therefore the ancient Egyptians thought that the world was static and unchanging. In this context, time was logically thought of as cyclic, consisting of a succession of eternally repeating phases. There was therefore little sense of past and future, evolution, or history.

Egyptian mythology asserts that in the beginning there was Nun, a 'non-being' which had the potential for life but was not alive as such. Then Nun gave life to Atum, a 'complete being' and Lord of the Universe. Atum manifested himself as Ra, the Sun god: a radiant dawn that filled space with the light of life. Then Atum generated the first pair of gods, Shu (male, and representing air or light) and Tefnut (female, and representing moisture). In turn this couple begot Geb, the Earth god, and Nut, the sky goddess, and finally these had four children: Osiris, Isis, Seth, and Nephthys. The tenth god was Horus, a heavenly divinity usually represented with the features of a falcon, who was the son of Osiris and his heir to the kingdom on Earth. An important point is that the gods created the world with everything in it already organised in a regular, permanent pattern. To use more modern terms, this is a static world.

When it comes to the physical universe, ancient Egyptians described the sky as a roof placed over the world, which was supported by four columns placed at the four cardinal points. The Earth's shape was a flat rectangle, longer from north to south than from east to west, and having (rather unsurprisingly) the Nile as its centre. Towards the South there was another river, this one in the sky, supported by mountains. It was on this river that the Sun god made his daily trip. Every evening the goddess Nut swallowed the sun; it then travelled through her body during the night, and she gave birth to it again every morning. However, this unchanging cosmic balance, with its regular and indeed predictable recurrence of the seasonal phenomena, did not occur spontaneously: its stability could only be ensured by a permanent and deliberate control. On Earth this was the function of the pharaoh: his main role was to ensure that the Sun would rise in the east and set in the west every day.

This idea of a stable and regularly repeating pattern of events, which was manifest in everyday life, naturally led to sense of security from the risk of change and decay. Thus there was no motivation for creativity or progress, which today is manifest in the fact that Egyptian art changed relatively little over a period of 3000 years. Another manifestation of this frame of mind is that the years were not counted in a linear sequence. Instead, they were counted with reference to the reign of each particular pharaoh: the counting would be reset to one when each pharaoh took over the throne. Today we have long (indeed almost complete) royal lists, but a lack of precisely dated events because it is difficult to match this list to a standard calendar. This is manifest in the surviving works of the historian Manetho—whose work actually dates from the Ptolemaic Kingdom of Egypt.

On the practical side, the Egyptians did make a key contribution to horology, the science of calendars and time measurements. They created a civil calendar which consisted of 12 months, each containing 30 days, and, clearly noticing that such a 360 day cycle did not stay aligned with the seasons and the Nile flooding cycle, supplemented it with five additional days at the end (known as the 'five days beyond the year'), making a total of 365. Historical evidence shows that this was in use as early as 2800 BCE. This is remarkable in being the first non-astronomical calendar.

This civil calendar had a purely empirical origin, built up by regularly observing, recording, and in the end averaging the time intervals between successive arrivals of the Nile flood. The year was divided into three seasons of 4 months each: Akhet (the Flood Season), Peret (the Growth Season) and Shemu (the Dry Season). Similarly, each month was divided into three 'weeks' of 10 days each, known as decans (which is actually a later Greek term)—a choice possibly related to the fact that there were ten main gods in the Egyptian pantheon.

To each decan period was associated a star or group of stars (themselves known as decans) which would rise in the east at dawn, just before the Sun itself (and after a period when they were invisible, being hidden behind the Sun). Such stars are said to be in heliacal rising. Just as the position of the Sun in the sky can be used to identify the time of day, the stars that are seen in heliacal rising on a given day (or, analogously, heliacal setting) can be used to identify the days of the year.

It is thought that, before this calendar, there was an earlier lunar-based calendar, with months beginning on the first day on which the old crescent was no longer visible in the east at dawn. This would be commensurate with the fact that their days ran from sunrise to sunrise—a convention that lasted until Hellenistic times.

Initially, the Egyptians did not realise that the astronomical year, and thus the cycle of the seasons, does not consist of exactly 365 days, but is in fact slightly longer, by almost 6 h. This difference was eventually recognised and another calendar was then introduced (probably around 2773 BCE) to track astronomical phenomena more closely. The crucial step is thought to have been the realisation that the rising of the Nile coincided with the heliacal rising of the star known as Sopdet to the Egyptians, Sothis to the Greeks, and Sirius to us—that is, the brightest star in the night sky. This fortunate coincidence was surely seen as a meaningful omen and provided the natural beginning of the Egyptian year in the 'Sothic' calendar. The Sothic calendar kept pace with the seasons, while the civil calendar did not; the two move steadily apart, and only coincide at intervals of 1460 years.

The low levels of cloud cover not only enabled easy observation of the night sky but also made the Sun a convenient clock. Thus the fact that the earliest known solar clock comes from Egypt is not particularly surprising. It has been dated to about 1500 BCE. It is also known that the pharaoh Tuthmosis III (ca. 1450 BCE) referred to the hour indicated by the Sun's shadow at a particularly important point of one of his military campaigns in Asia, which indicates that portable solar clocks were also in use.

In order to measure time at night (or in general, when the Sun was not available), the Egyptians also invented the water clock, which we now know as the 'clepsydra', as the Greeks later called it. Both the obvious main types are known to have been developed and used, with water flowing out of or into a graduated vessel, and clepsydrae were also used by the Greeks and Romans. Finally, the Egyptians used a third instrument to observe the transits of the relevant stars across the meridian. This was a set of two plumb lines, which they called the 'Merkhet'. The principle behind this is the same as that of the heliacal risings: on a given night different stars (specifically, different decans) transit the meridian at different times, and for each decan the transit time varies according to the day of the year.

It is also worthy of note that our modern division of the day into 24 h ultimately stems from Egypt. That said, a subtle but crucial difference is that these hours were not of equal length. Instead, at all times of the year the periods of daylight and darkness were each separately divided into a period of 12 h. The end of the night was determined by the heliacal rising of the appropriate decan, as has already been explained. These initial divisions of the day and night into separate periods of 12 h were subsequently replaced, in Hellenistic and Roman times, by a single period of 24 'seasonal' hours of the full day. Thus the actual length of 1 h varied according to the day of the year and, moreover, on each day except at the equinoxes the actual length of a day-time hour and a night-time hour were different. In a place like Egypt which is close to the Equator these daily differences would have been small, but they would of course become much greater if the same concept were to be applied at higher latitudes. In antiquity, only the Hellenistic astronomers regularly used hours of equal length, and naturally those hours were chosen to be the same as the seasonal hours on the day of the Spring equinox. Gradually this uniform definition became the norm.

Since, following Babylonian practice, all Egyptian astronomical computations involving fractions were done using the convenient sexagesimal system (rather than our current decimal system), those Egyptian hours were then divided by the astronomers into 60 *first* small divisions (*pars minuta prima*, in Latin), or minutes, and each of these was in turn subdivided into 60 *second* small divisions (*pars minuta secunda*, in Latin); these are the origins of our terms 'minute' and 'second'. Thus our modern-day convention for dividing up and subdividing the hours of the day is the result of a Hellenistic modification of an ancient Egyptian practice, combined with Babylonian mathematical conventions.

Interestingly, Egyptian mathematics used the following approximation for π :

$$\pi = 4\left(\frac{8}{9}\right)^2 = 3 + \frac{13}{81} \sim 3.1605, \tag{1.1}$$

while Babylonian mathematics used the cruder approximation $\pi = 3$ or sometimes

$$\pi = 3 + \frac{1}{8} = 3.125, \tag{1.2}$$

and the approximation $\pi = 3$ can also be found in the Old Testament.

1.2.2 Mesopotamia

At the risk of oversimplifying, one could say that while Egypt excelled more in the arts, Mesopotamia excelled more in technology and science. Another important difference is that while the Egyptian civilisation developed in a fairly smooth way over a period of several millennia (the most dramatic difference being the final Ptolemaic period following the conquest by Alexander the Great), many different civilisations flourished in Mesopotamia and the surrounding area over the corresponding millennia.

Specifically, one should mention the Sumerians (ca. 3000–2000 BCE), Babylonians (2000–200 BCE), Hurrians and Hittites (1700–1300 BCE), and Assyrians (1400–600 BCE). Of particular interest for our present discussion is the period of the Persian empire, from 539 to 331 BCE. This is effectively when astronomy was developed by Babylonian scholars. It was then quickly adopted by the Greeks, who eventually surpassed them at the beginning of the common era.

The earliest studies of astronomical science in Mesopotamia probably date from around the end of the fourth millennium BCE. There is considerable uncertainty in this statement, since the first written texts to have survived until the present, from the Babylonians, and subsequently from the Assyrians, only date back to the second millennium. The most famous of the Babylonian myths is known as the Poem of Creation (*Enuma elish*). This describes the sky as one half of the body of the goddess Tiamat, who was defeated by Marduk following a terrible cosmic fight. Since the heavens were considered sacred, they were thought to provide important omens, and the careful interpretation of celestial phenomena enabled priests to predict both natural and political events. It also compelled rulers to pay attention to these forecasts and react appropriately whenever necessary.

The Babylonian pantheon included Shanash (the solar disk at dawn or dusk) which was humanity's benefactor, while Nergal (the solar disk at noon) was an evil and destructive force, and Sin (the Moon) was the god of wisdom. The planet's names were Marduk (Jupiter), Ninib (Saturn), Nabu (Mercury), Ishtar (Venus), and Nergal (Mars). Surviving records of Moon risings date from about 1800 BCE, and a set of records of risings and settings of Venus over a period of 21 years (1702–1681 BCE) also survives. A particularly remarkable astronomical catalogue, known as the Mul.apin (see Fig. 1.3), dates from 686 BCE and survives in several copies, but is believed to have been compiled in the Assyrian period, in about 1000 BCE. It lists the names of 66 stars and constellations, as well as information on risings, settings, and culminations. The broad astronomical content and significance of the Mul.apin was only identified in recent times, in 1880.

The Babylonians apparently believed the Earth to be a large circular plate, surrounded by a river which no human could ever cross and beyond which there was an impassable mountain barrier. It was this mountain barrier that supported the vault of heaven, which was made of a very strong metal. Everything was resting on a cosmic sea. There was also a tunnel in the northern mountains that opened to outer space and also connected two doors, one of which was towards the east and the other towards the west. Every day, the Sun came out through the eastern door, travelled below the metallic heavens, and finally exited through the western door, spending the night in the tunnel. As you can see there are interesting similarities—but also differences—with respect to the Egyptian version.

Fig. 1.3 One of the Mul.apin tablets. Its actual height is 8.4 cm (Public domain image)

Another crucial difference between Mesopotamia and Egypt pertains to the local environment and climate: the Tigris and Euphrates were far less predictable than the Nile. While the Nile would only very rarely bring unexpected drought or flooding, the climate in Mesopotamia was far more variable. Strong winds, torrential rains, and devastating floods were frequent and would appear with little or no warning—let alone control. In this context, the regularity of the astronomical cycles must have been particularly noticeable—as well as reassuring—in an uncertain and violent world.

Since the Sun, Moon, and planets were seen as gods (or at least visible manifestations of gods), the king and his counsellors must have watched avidly for portents that could be interpreted in terms of the intentions of these gods, so that disasters might be foreseen and possibly avoided or mitigated. The underlying assumption was that a counterpart in human events existed for every observable celestial phenomenon.

This was therefore the motivation that led priests to make careful and systematic observations of the heavenly bodies. The previously mentioned surviving evidence shows that this was already being done on a large scale in the eighteenth century BCE, but there is further evidence suggesting that particular phenomena (such as lunar eclipses) may have been regarded as ominous well before that epoch.

Gradually, the Babylonian civilisation acquired a remarkably accurate knowledge of the periodicities of the heavenly bodies: even relatively inaccurate observations can lead to accurate estimates provided one has a sufficiently long time series. Having accumulated enough data, they could use these periodicities to predict future positions of these bodies, and in particular predict the occurrence of specific phenomena such as lunar eclipses. Nevertheless, it is worth bearing in mind that this knowledge was limited to a small elite (the priesthood), and there was, as far as we know, no attempt to formulate any underlying theory of celestial motions (either geometrical or any other type).

Babylonian calculations were entirely empirical: they could predict when phenomena such as eclipses would occur, but did not (as far as we know) worry about why they occurred. Their algorithms were a collection of empirical facts without a theory—a bit like a phone book. On the other hand, the Greeks did worry. One could therefore say that astronomy was born in Babylon, but cosmology was only born later in Greece.

At this point, astronomy and astrology were born together. The predictions of what is now known as 'judicial' astrology applied to the royal court and the state as a whole, and not to ordinary individuals. This is to be contrasted with so-called 'horoscopic' astrology, which is closer to its modern remnant. The basis of horoscopic astrology is the assumption that the positions of the planets at the time of birth will determine the fate of the individual. This version only developed during the Persian domination.

In order to be able to cast such horoscopes one must have information about the positions of the planets for a given date, and this date can of course be one for which no observations have been recorded. Thus any such endeavour needs general methods for computing the positions of the planets on any specified day, possibly by interpolating from observations before and after that day. The oldest known system of Babylonian planetary algorithms (to use the modern term) satisfying these requirements is thought to have been developed not earlier than 500 BCE. Some surviving tablets from the third century BC describe such methods for phenomena including the new Moon, the opposition and conjunction of the Sun and Moon, and eclipses.

Another motivation for studying the heavens was the development of a reliable calendar. The Babylonians had a lunar calendar, and the month began when the new lunar crescent was visible for the first time, towards the west, shortly after sunset and soon after the new Moon. Thus the Babylonian day began in the evening. A lunar month that is counted in this way must contain a whole number of days. The average lunar synodic month (the interval between successive new Moons) is about 29.53 days, but since Earth's orbit around the Sun and the Moon's orbit around the Earth are both elliptical, the actual duration of each cycle can vary significantly, approximately from 29.18 to 29.93 days. Thus a lunar month defined as above will sometimes have 29 and sometimes 30 days. It follows that a lunar year containing 12 months will have about 354 days, which is significantly less than the solar year, so some correction will be needed if one wishes to keep this calendar aligned with the seasons.

A simple solution is to insert a thirteenth month from time to time—a process known as an intercalation. As far as we know, there was initially no regular system for this intercalation, which was presumably only used when the misalignment of the calendar was noticed. Starting sometime in the fifth century BCE, a regular system was introduced whereby 7 of these months had to be inserted at fixed intervals in a cycle of 19 years. This was based on the realisation that 19 solar years are, to a very good approximation, equal to 235 lunar months. This is usually known as the Metonic cycle, after the Athenian astronomer Meton who introduced it in Athens in 432 BCE, together with his disciple Euctemon. Whether this cycle was discovered first in Babylon and then imported into Athens or independently by the Babylonians and by Meton is uncertain, although it has been suggested that the Babylonians were using it as early as 499 BCE.

What we now call the zodiac (from the Greek *zodion*, or 'little animal'), the belt around the celestial sphere in which the Sun, Moon, and planets move, was also invented around that time. The 12 zodiacal signs, each of which was crossed by the Sun in about 1 month, are known to have been in use soon after 500 BCE. This division of the sky corresponded to a division of the circle, leading to the habit of dividing a full two-dimensional rotation into 360°, so that the Sun travels round the celestial sphere by about 1° per day. This quantification was an important step in mathematical astronomy, which was also helped by the fact that the sexagesimal system (probably developed in the third millenium BCE by the Sumerians and then adopted by the Babylonians) is computationally easy, since many fractions simplify in this system. Thus the development of such a system must have been the result of a combination of practical mathematical convenience and religious and astrological reasoning.

Eventually, the 19 year luni-solar cycle became the cornerstone of the Jewish and Christian calendars, since it provided a computationally simple method for establishing the dates of new Moons for religious purposes, including the crucial question, in the Christian calendar, of determining the date of Easter (since that affects most of the rest of the annual religious calendar).

The Babylonians also paid particular attention to the dates, separated by intervals of approximately 7 days (though on average slightly more than this), associated with each phase of the Moon. Each of these phases ended with an 'evil day' on which specific prohibitions were enforced, with the goal of appeasing the gods. This habit influenced the Jews, who in their turn influenced the early Christians and eventually ourselves. This is the ultimate origin of our 7-day week and of the traditional restrictions which apply to activities on a particular 'holy day'—in our case Sunday—and this origin can thus be traced back to the Babylonians.

The Hebrews borrowed much of Babylonian culture during their captivity in Babylon in the sixth century BCE. In particular, they were influenced by the Sumerians and Babylonians in the way they measured time, so they also had a lunar calendar with the month beginning in the Babylonian way. Intercalary months were also used to keep a reasonable agreement with the Sun. Both the new Moon and the full Moon were regarded by the Hebrews as being of religious significance. In particular, the timing of Passover was determined by the first full Moon occurring on or after the spring equinox. In fact, cuneiform records indicate that the Babylonian word *Shabbatum* actually meant full-moon day, and this is a remnant of what must have been the primary meaning of the Hebrew term 'Sabbath'.

Although the Hebrew 7-day week ending with the Sabbath resembles the aforementioned Babylonian period of approximately 7 days ending with an 'evil day', there are two important differences. The latter cycle was always directly correlated with the Moon's phases (implying that the evil days did not always occur at strictly 7-day intervals), whereas the Hebrew week was not, but was continued from month to month and year to year at 7-day intervals, regardless of the Moon's phase. Moreover, the Babylonian evil day was only observed by the court elite (the king, priests, and physicians), whereas the Sabbath was observed by the whole nation. Decoupling the week from the lunar month and employing it as a further unit for the calendar was an innovation which it seems can be attributed to the Hebrew people. Because of its origin, Christianity also inherited the Hebrew week.

1.3 Clocks and Calendars

Imagine how our lifestyle would be, and the development of civilisation would have been, if the sky was completely and permanently covered with thick clouds. It would be much harder to tell the time, we might have no calendars, and navigation and exploration of the planet would have been correspondingly harder—and also slower. Even disregarding the obvious effects on agriculture and other areas, the emergence of civilisation would at best have been significantly delayed.

We are so accustomed to the idea of time and its divisions, and the role they play in our lives, that we tend to have no idea that our own society's view is historically peculiar, and indeed we are more obsessed with time than almost all other societies before ours—the only possible exception could arguably be the Maya. Think of how many times, when you wake up, your first reflex is to look at a clock, and how often you check what time it is using the many clocks and watches that you carry or can find around you. Many of us carry diaries not to record what we have done (as would be the case just one or two generations ago) but to make sure we are at the right place at the right time.

An inner sense of time is one of the fundamental characteristics of human experience, but interestingly whether or not we have a special sense of time (as we have of sight, hearing, and so on) is an open question. If a 'sense of time' is understood as an awareness of duration, it is clear that this will depend on several internal and external factors. The most important such factor is clearly age: as we get older, time measured by the clock and the calendar appears to us to pass ever more rapidly.

Most people tend to feel that time goes on forever and is completely unaffected by anything else. This is effectively the Newtonian view of time—part of the static, pre-existing, and universal system of rulers and clocks that we can imagine permeate the Universe. In this view, if all activity were suddenly to stop, time would still go on unaffected. Although this is a natural view within our classical intuition, it is demonstrably wrong: one can show, experimentally, that this is not how the Universe works. In particular, as Einstein has shown (and we will discuss later on), that system of universal rulers and clocks does not exist. Our linear concept of time is also exceptional: most civilisations before ours have tended to regard time as essentially cyclic in nature.

To the extent that one can divide historical developments into simple 'before' and 'after' periods, the defining moment in the history of the measurement of time was the invention of the mechanical clock in Western Europe, towards the end of the thirteenth century. This was not done for practical reasons, but for theological ones: monks needed to know at which times they should pray.

The reason this invention was conceptually revolutionary (and not merely an incremental improvement) is that time can, equally intuitively, be thought of as a continuous quantity, as may be measured by the flow of water or sand, the burning of a candle, or the gradual motion of the shadow of a stick. The mechanical clock, relying as it does on the escapement mechanism (an oscillating device which periodically interrupts an otherwise continuous process such as the falling of a weight), forces one to discretise time, and hence to think of it as a succession of individual units, which can be added up and combined into arbitrarily long periods.

Interestingly, for many people not only is time universal and absolute, but so is the way we measure it. Recent history records several objections to the introduction of Summer Time, on what are effectively theological grounds. Moreover, when the Julian calendar was replaced by the Gregorian calendar (introduced by Pope Gregory XIII in Western and Catholic Europe in 1582, and later elsewhere), which was done by suppressing some calendar days (the day after 4 October was 15 October), many people thought that they were losing a few days from their lives.

Many people are also surprised when they first realise that there are different time zones across the planet, and even an International Date Line. The first is now closer to our everyday experience, since long-distance flights are increasingly common—as are their physiological effects, in the form of jet-lag. While the International Date Line has no associated physiological effects, it is perhaps less intuitive: one must move the calendar, not the clock, forwards or backwards: by crossing this line, one either loses or gains a whole day in the calendar because of the time difference of 24 h between the east and the west of the line. Therefore, by doing so we will experience a week with either 6 or 8 days.

What is happening is that, when we go round the world towards the east, the number of days we count will be one more than the number of days counted by someone who stays at the point of departure and arrival. The reason is that by traveling towards the east, each day of the journey will be somewhat shorter than 24 h. The opposite occurs when we travel towards the west. This is the key to the plot of Jules Verne's story 'Around the World in 80 Days'. When the hero came to the end of his journey eastwards he thought he had taken more than 80 days, until he realised that he had forgotten to put his calendar back when crossing the Date Line. An earlier example involving travel towards the west is provided by the crew in Magalhães's circumnavigation of the world: they were puzzled when they reached Cape Verde in 1522 on what they thought was a Wednesday, only to find that the inhabitants were treating it as a Thursday.

The way we measure time is obviously based on local astronomical phenomena. The rotation of the Earth gives us our day, while the Earth's motion around the Sun gives us our year. On the other hand, the way we subdivide the day into hours, minutes, and seconds is conventional, as was already discussed. Nowadays, the measurement of time is so precise, thanks to atomic clocks, that small variations in the Earth's motion are noticeable. Thus our exquisite modern methods of time measurement are the reason for the frequent introduction of leap seconds.

Almost every possible convention has been adopted by someone for the beginning of the day. As we already mentioned dawn was chosen by the ancient Egyptians, while sunset was chosen by the Babylonians, and thence Jews and Muslims. The Romans initially chose sunrise but later midnight. Dawn was also used in Western Europe before mechanical clocks, but later midnight was chosen as the beginning of the civil day. Astronomers naturally found it more convenient to choose midday (to avoid a change of day during a night of observations), and this convention was only abandoned on 1 January 1925 when, by international agreement, the astronomical day was made to coincide with the civil day.

The stars, particularly their rising and setting, can also provide accurate clocks. Sidereal time is a measure of the position of the Earth in its rotation around its axis, which is very close (although, strictly speaking, not identical) to the motion relative to the stars. During the time taken for the Earth to complete one rotation around its axis (a sidereal day, of 23 h 56 min 4.09 s) the Earth moves along its orbit around the Sun. Therefore after a sidereal day, the Earth still needs to rotate a little bit more for the Sun to reach, say, its highest point on the sky (towards the south in the northern hemisphere), which is the basis of the definition of a solar day of 24 h. A solar day is therefore around 4 min longer than a sidereal day.

By comparison to the Earth–Sun distance, the stars are so far away that the Earth's movement along its orbit makes a generally negligible difference to their apparent direction, and therefore they return to their highest point in a sidereal day. Overall there is one less solar day per year than there are sidereal days. As has already been mentioned, heliacal risings and settings vary throughout the year and can form the basis of a calendar.

Although the stars can provide a calendar that will necessarily be aligned with the seasons, they do not provide a neat division of the year into convenient smaller parts. The Moon was therefore used for this purpose. Indeed the periodicity of a month is easier to identify than that of the year, but unfortunately the two do not match smoothly since the solar period is not a convenient multiple of the lunar period. Thus if one has a calendar with lunar months, each of which begins with the observation of the New Moon, keeping the calendar aligned with the seasons will require an ad hoc solution such as the intercalation of an additional month (as we have already seen). In fact, many cultures developed two calendars: an agricultural one based on the Sun, and a sacred one, based (at least in part) on the Moon. The months in our calendar no longer have any connection with the Moon, but are an effectively arbitrary way of dividing the solar year into 12 parts.

Our modern calendar has a long history, starting in Egypt and Babylon but ending in Rome. Originally, the Roman calendar began on 1 March, which is easily seen today in the names we give to the months from September to December—the seventh to tenth months of the year, but around the year 158 BCE the date of 1 January was chosen. Our calendar is a politically modified version of the one introduced by Julius Caesar on 1 January 45 BC, and since named after him. Before this, the Romans had a lunar calendar with an intercalary month every second year. This month was called Mercedonius, ordinarily lasting either 22 or 23 days. But the length of this month was not determined by any precise rule, so the Collegium Pontificum (the body that presided over religious matters and the management of the calendar) could—and indeed often would—manipulate the calendar for political ends. The result was that by the time of Julius Caesar the calendar was about 3 months out of phase with the seasons.

A two-step solution was proposed to Julius Caeser by the Alexandrian astronomer Sosigenes. The first step was to synchronise the calendar with the seasons by extending the year 46 BC to a total of 445 days—this became known as 'the year of confusion'. The second and longer term step was to abolish the lunar calendar and adopt a solar calendar. He adopted an Egyptian-type year of 365 1/4 days, with ordinary years of 365 days and a leap year of 366 days every fourth year. The months of January, March, May, July, September, and November (in other words, the odd-numbered months) should each have 31 days, while the even-numbered months should have 30 days, except February, which should normally have 29, but in a leap year would have 30.

Following the death of Julius Caesar, his birth month, formerly known as Quinctilis (the fifth month) was renamed Julius in his honour. In 7 BCE it was similarly decided to honour Octavian (by then called Augustus) by renaming his birth month, formerly known as Sextilis, after him. Political wisdom demanded that his month (August) should have the same number of days as July. Therefore a day had to be removed from February and transferred to August. To avoid having 3 consecutive months of 31 days, September and November then lost a day and were reduced to 30 days, while October and December gained a day and ended up with 31. Thus to honour the first of the Roman emperors, a rather neat arrangement was turned into a much more obscure one that many people still find difficult to remember.

Sometime at the beginning of the common era, the 7-day week became popular in Rome, although it did not achieve official status until Sunday was designated a holy day in the time of the Emperor Constantine. Nevertheless, the names the Romans gave to the days of the week can still be identified in most modern Western languages. It was popular, by astrological tradition, to associate the days in the 7-day week with the different days named after the planets, in the following order: Saturn, Sun, Moon, Mars, Mercury, Jupiter, Venus. This is the origin of our modern names for the days of the week. Still, the order may seem strange since it does not correspond to the order in which they were supposed to be placed relative to the Earth. As we will see in the following chapter, the classical Greek order was: Saturn, Jupiter, Mars, Sun, Venus, Mercury, Moon.

As it turns out, the explanation is simple—and again, astrological. The planets were believed to rule the hours of the days as well as the days of the week, and each day was associated with the planet that ruled its first hour. The first day of the Roman week was Saturday, and its first hour was therefore ruled by Saturn, who also ruled the 8th, 15th, and 22nd hours. The 23rd was then assigned to Jupiter, the 24th to Mars, and the first of the next day to the Sun, which thus ruled Sunday, and so on throughout the week. This so-called 'planetary week' is used in many modern languages, sometimes in the local mythology rather than the Greco-Roman one. This is the case in English, as one can realise by noticing that Tiw, Woden, Thor, and Freya are the Norse names of the corresponding gods. The languages of Western Europe and India typically use planetary or deistic names, whereas those of Eastern Europe and the Middle East are ordinal (just numbering the days of the week). Portuguese is an exception, and for some unexplained reason it is ordinal.

1.4   Science in the Modern World

The original role of science and scientists was simply to provide an account of the world as it is. Although the term 'scientist' did not exist at the time of Galileo or Newton, this is how they would have seen their own roles. But throughout the twentieth century it became increasingly clear that doing science also entails a social responsibility.

Today, as has already been mentioned, this social responsibility is all the more compelling given our society's reliance on science and technology and the widespread scientific illiteracy of the population. As James Burke so succinctly put it, never have so many people understood so little about so much.

Scientific literacy can be defined as a minimum level of understanding of science and technology necessary for any member of a modern society. This has two main components. The first is more abstract, consisting of a conceptual understanding of the principles of the scientific method, including knowing what a scientific hypothesis is, what an experiment is, and what constitutes evidence in favor of or refuting a particular hypothesis, and so on. The second is more practical and involves a basic understanding of key concepts. A partial list of these could include, for example, the following: atom, cell, DNA, electron, evolution, gravity, greenhouse effect, LASER, light, molecule, natural selection, neuron, nuclear energy, photon, proton, star, and stem cell.

It is important to note that scientific literacy does not require or assume any advanced or formal knowledge. It is simply a repository of basic knowledge that any active member of a modern society should have in order to be actively involved in it and able to participate in debates over society's main challenges.

Quantifying the degree of scientific literacy of a population is not easy, so most studies focus on functional literacy, which quantifies how well someone can make use of scientific literacy in practical daily situations. For example, you may be given a box with antibiotics, be told to read the instructions therein, and then be asked what dose you should be taking and how frequently. Such studies indicate that the highest degree of scientific literacy is found in Scandinavian countries (at the level of around 30% or so, with Sweden usually ranked first), while in southern Europe the level is below 10%.

A seemingly surprising result is that the level of scientific literacy in the USA is also close to 30%. This is surprising given that the USA is rife with creationism, global warming denial, anti-vaxxers, and other such nonsense. But there is actually a good explanation: this is roughly the percentage of people in the USA with a university degree. In American universities all degrees, whether or not they are science based, have a minimum requirement of credits in science topics (there are some famous courses such as Astronomy for Poets), and in such courses this basic knowledge is absorbed. This number makes me quite optimistic: if we want to increase the level of scientific literacy in a country like Portugal by a factor of three or four, I know how to do it: just make it compulsory for everyone doing a university degree to take a couple of scientific courses of their choice. Granted, this will take one generation, but it is entirely achievable.

The mathematical equivalent of the concept of scientific illiteracy is called innumeracy. This is significantly more common than illiteracy and—more worryingly—it is also socially accepted. Nobody in a modern society would publicly admit that they don't know how to use a computer, or send an e-mail, or speak English, but many people are perfectly happy to admit that they were never good at maths. This social acceptance is deeply problematic, and it implies that it will be a much harder problem to solve.

The prevalence of innumeracy in the media is obvious, and it hardly needs to be pointed out. But it is worth mentioning a few of its symptoms, which are of several different kinds. Some of these, like having difficulties in dealing with very large and very small numbers, confusing precision and accuracy, are basic mistakes due to inexperience or ignorance and could easily be overcome with some comparatively simple training, starting in secondary school.

Then there are issues of statistics, such as confusing correlation with cause and effect, or misinterpreting conditional probabilities, which may require a more significant amount of training. The first of these is particularly relevant in the current era of data mining: with the unprecedented amount of computing power at our disposal we can sieve through huge amounts of data, and it is a virtual statistical certainty that, with a large enough set of variables, there will be correlations between them without their being related in any possible way.

Finally, there are deeper biases in the way our brains work, such as overestimating insignificant or exotic risks (especially when compared to greater but more diffuse risks), or identifying patterns in random events. These have also been identified in other primates and seem therefore to have been imprinted in our brains long ago in evolutionary terms, so the only thing one can do is recognise that we are vulnerable to these biases and do our best to be alert and avoid being fooled by them.

A classical example of our difficulty in evaluating and comparing risks is the idea that many people have that flying is more dangerous than driving a car. In fact, the most dangerous part of a typical trip involving a commercial flight is (by far) the drive to or from the airport. Even if terrorist groups crashed 1 commercial plane per week, a typical passenger would need to fly 23 times a month to have a greater risk of dying than they would on the road.

Why does this happen? One reason is that whenever a plane crashes and kills everyone on board it makes news headlines for several days. Overall much greater numbers of people die on the road, but they do so one or two at a time, and these diffuse deaths don't generate the same headlines, so we never hear of them unless we happen to be personally affected by them. Thus people can more readily remember examples of plane crashes than road traffic accidents and—due to what psychologists call the availability heuristic—take a mental shortcut to assume that plane crashes must be common and road traffic accidents must be rare, whence flying must be more dangerous than driving.

A similarly biased perception, for exactly the same reasons, occurs for the risks of energy production in nuclear power plants versus thermal power plants burning coal. In fact, the production of electricity in thermal power plants causes 100,000 deaths per year (due to cancer, respiratory diseases, and many other problems); if all these thermal power plants were replaced by nuclear power plants, one Chernobyl-like accident would have to occur every 3 weeks to cause as many deaths. This biased public perception of the supposed risks of nuclear power plants is a major obstacle limiting our ability to effectively tackle global warming.

Of course, another telling example of scientific illiteracy is the modern-day belief in astrology. The origins of the current version of astrology have already been mentioned (actually they coincide with those of astronomy), and its principles are seemingly simple: a person's character, daily life, and ultimate destiny can be understood from the positions of the Sun, the Moon, and the planets of the Solar System at the moment of their birth.

When astrology arose in Babylon, our world vision was dominated by magic and superstition, and for the reasons we have already discussed, trying to understand the patterns of heavenly bodies was often a matter of survival. Thus people would systematically observe the heavens to search for signs of what the gods would be doing next. In this frame of mind, a system that connected the planets and the zodiacal constellations in which we observe them with the everyday life of ordinary humans was psychologically appealing and reassuring.

Astrology is a psychologically clever way of blaming something (or someone) else for one's problems or misfortunes: something beyond our control can't be our own fault. In this way people are failing to accept responsibility for their own lives. At the same time astrology also gives people a sense of apparent control over their destiny: although we don't know what the future will bring to us, it is reassuring to know that someone does know. The final piece of the puzzle is easy: people have a tendency to accept a vague and completely general personality description as uniquely applicable to themselves, without realising that it could be applicable to everyone else. This is the prime example of what psychologists call the Forer effect.

In case there were any doubts, many double-blind tests have conclusively shown that astrology simply does not work, and despite their many claims, astrologers can't make any predictions. After all, as has already been mentioned, we don't need to know how something works to determine whether or not it works.

Recalling our discussion of the scientific method, if astrology were a science its predictions should have gradually converged into a consensus theory, but this has never happened. On the contrary, not only there are multiple astrological schools (apart from the Western one, China and India have their own, very different ones), but even within each such school there are almost as many versions and interpretations as there are astrologers. Scientific ideas progressively converge as they are tested against a growing amount of experimental evidence. On the other hand, systems based on superstition or personal belief not only do not converge, but in fact tend to diverge.

The planet Uranus was only discovered in 1781, and Neptune in 1846. The history of the discovery of Neptune (which we will discuss later) also contains an important lesson: astronomers were able to predict its position mathematically before it was discovered. If, as astrologers claim, the influence of all major bodies in the Solar System must be taken into account, surely all pre-Neptune (or should we say pre-Pluto?) horoscopes must be wrong. If there was any scientific basis to astrology, one would have expected that astrologers would notice the inaccuracy of their predictions and deduce from it the presence of further planets at specific positions. But such a process is impossible in an astrological framework.

A final historical point is that the ecliptic constellations used by the astrologers, and the dates on which the Sun is supposed to be in each of them (which supposedly determine the person's star sign), were defined 2500 years ago, and due to various dynamical phenomena, such as the precession of the equinoxes, the Sun gradually moves about in the sky and the interval during which it is in each constellation gradually changes, so the old correspondence between constellations and dates is simply wrong today. Bearing in mind that the definition of the constellations is purely the result of a choice—see Fig. 1.4—there are in fact 13 constellations on the ecliptic, and at present the Sun is in each of them on the days shown in Table 1.1.

Note that the intervals are longer for some constellations than for others— the constellations are not of equal size, though once again the borders between them are conventional. Most historians of astronomy believe that the very first constellation makers lived around 2700 BCE (give or take a

Fig. 1.4 A celestial map from 1670, by the Dutch cartographer Frederik de Wit (Public domain image)

Table 1.1 The actual constellations of the ecliptic, and the dates in which the Sun can be seen in the sky in the direction of each of them

Constellation	Start	End
Sagittarius	18 December	18 January
Capricornus	19 January	15 February
Aquarius	16 February	11 March
Pisces	12 March	18 April
Aries	19 April	13 May
Taurus	14 May	19 June
Gemini	20 June	20 July
Cancer	21 July	9 August
Leo	10 August	15 September
Virgo	16 September	30 October
Libra	31 October	22 November
Scorpius	23 November	29 November
Ophiuchus	30 November	17 December

Note that given the existence of leap years, the limits may vary by about 1 day from year to year couple of centuries). The first known cuneiform texts which refer to 12 equal constellations are from the fifth century BCE. The most ancient Babylonian texts (e.g., from around 700 BCE) contain different and more irregularly-sized constellations. The dates that astrologers quote were reasonably accurate over 2000 years ago, when astrology was first created. By now, we are almost a whole sign away from what today's astrologers claim.

Note also that there are currently 13 constellations, not 12, the thirteenth being Ophiuchus. Interestingly, some early Greek horoscopes did have 13 constellations, but this was eventually reduced to 12 because the number 13 was considered an ill-omen. It has also been suggested that the reason for the final choice of the number 12 is that the solar year is approximately composed of 12 lunar months, so it was natural to divide its annual path into 12 parts. Moreover, there were 12 ruling gods in Greek mythology.

The solution eventually adopted for reducing the number of constellations from 13 to 12 (eliminating Ophiuchus) was not unique. One alternative was to keep Ophiuchus but enlarge the constellation of Scorpius, which included the stars of Libra, while a second alternative was to include the stars of Libra in Virgo. Note that Libra is unusual among the zodiacal constellations in that it is the only non-living being.

The example of Ophiuchus shows that, over longer time periods, constellations can come in or out of this list. Looking further into the future, two other constellations will become part of the list: from May 5451, someone could be of the star sign Cetus, and from September 6381, they could be of the star sign Orion.

The above constellations are usually called the zodiac by astrologers. In fact, a more rigorous term would be the ecliptic constellations—these are the ones through which the Sun is seen to move. In addition to the Sun, the planets are also supposed to move about on the sky in one of these constellations. In fact, the situation is somewhat more complex. Because the orbits of the other planets are not exactly in the same plane as the Earth's orbit, they can in fact be found at times in other constellations. If one defines the zodiac more broadly as the set of all constellations in the sky where planets can possibly be found, it consists of the following, in addition to the ecliptic constellations: Auriga, Bootes, Canis Minor, Cetus, Coma Berenices, Corvus, Crater, Eridanus, Hydra, Leo Minor, Orion, Pegasus, Scutum, Serpens, and Sextans.

There is, of course, no complete solution …But we can do something. The chief means open to us is education—education mainly in primary and secondary schools, but also in colleges and universities. There is no excuse for letting another generation be as vastly ignorant, or as devoid of understanding and sympathy, as we are ourselves.

Charles Percy Snow (1905–1980)

2

Classical Astronomy

In this chapter we focus on Classical and Hellenistic Greece, looking in detail at their astronomy (which was in part imported from Babylon and Egypt), but also at their cosmology and physics (which they pioneered). We will also discuss several Greek philosophical schools, which impacted these developments and led to what can fairly be called the first canonical cosmological model. Its conceptual structure is mainly due to Aristotle, while its technical details were provided by a number of mathematicians culminating with Ptolemy. This was our standard model of the Universe for more than 1500 years. We will also discuss how some of this knowledge was preserved in the period between the decline of the Greek civilisation and the Renaissance.

2.1 The Greek World View

As far as we know the Greeks were the first civilisation that thought of the night sky as displaying a set of phenomena that could and should be understood. This did not imply eliminating the gods as the primary causes of the phenomena, but it did mean assuming that the observed regularities of the phenomena were due to some set of impersonal laws. It then followed that if one knew these laws one could predict the behaviour of the objects. A standard example given to illustrate the point is that Thales of Miletus (ca. 624–546 BCE), the founder of the Ionian school and likely responsible for the introduction of Egyptian astronomy into Greece, is said to have predicted a solar eclipse, even though he had no concrete understanding of what caused

© Springer Nature Switzerland AG 2020
C. Martins, *The Universe Today*, Astronomers' Universe,
https://doi.org/10.1007/978-3-030-49632-6_2

the eclipses, or of the motions of the heavenly bodies. This eclipse took place on 28 May 585 BCE, and as it happens it occurred during a battle between the Lydians and the Medes. This fortunate coincidence, together with the fact that timings and other circumstances of eclipses can be very accurately calculated for the past as well as for the future, makes this the earliest historic event for which an exact date is available.

An interesting aspect of Greek mythology is the presence of three Fates: the task of these goddesses was that of weaving a rug which recorded all the affairs of both men and gods. Once something was weaved onto this rug, there was nothing that anybody could do to change it—even the gods were powerless to do so. This was the first appearance of the notion of an overarching mechanism or force which ruled everything in the Universe, even the gods themselves.

The sixth century BCE saw the emergence of various new philosophical schools of thought, sometimes alongside and sometimes mingled with mythological views. Philosophers started asking about the origin, ingredients, structure, and operation of the world. Is it made of only one thing, or of many? What is its shape, and its location? And what are its origins? Of course, the same questions had been addressed by mythology and religion. But the important point here is that the new explanations excluded the gods as direct causes—they are entirely naturalistic, emphasising the order, regularity, and predictability that was becoming ever more apparent.

It's not clear whether Thales claimed that everything was made from water, or only that it ultimately came from water. Anaximander (ca. 610–546 BCE) claimed that the primordial substance was *to apeiron* (the indefinite, in the sense of something boundless and indeterminate); this had no observable qualities of its own, but all observable phenomena could nevertheless be explained in terms of it. Anaximenes (ca. 585–525 BCE) had a preference for air, arguing that it could support the Earth like a leaf in the wind; in this case he definitely held that everything was made of air. He also provided a rational explanation for lightning and thunder (inspired by earlier ones due to Anaximander), and held that earthquakes occurred when the Earth was too dry or there was too much rain.

Most of the Greek philosophy of this epoch had as a primary aim that of distinguishing the external (or apparent) reality from the underlying (and presumably true) one. This stemmed from the view that the physical and observable world was only a partial—and as such incomplete and possibly misleading—reflection of a deeper underlying reality. In this view our senses are unreliable, and can only give rise to *gnome*, which can be translated as 'opinion' and is necessarily limited and superficial. One may think that a phenomenon happens in a certain way because one observes it to be so, or it seems be so, but the true underlying reality of the world may be different and possibly even beyond our senses. This true knowledge was called *episteme*. This dichotomy was most vigorously defended by Plato.

The Classical Greeks saw the past as a Golden Age of gods and heroes, and interpreted history as a gradual but unavoidable decline from this ideal state. This explains why, for the Greeks (unlike for the Persians, for example), there was no god of time. Only very late, in Hellenistic times, was a deity worshipped under the name 'Aion', but that was a sacred, eternal time which had very little relation to ordinary time. They called this ordinary time *Chronos*—and that always denoted a time interval, not time itself. It should be noted that the Greek concept of 'divine' referred to entities that were 'alive' (in the sense of being able to cause motion) and never died.

While it is often said that Herodotus and Thucydides are the founders of history, it is clear that their account of historical events is somewhat different from the modern one, especially when they had to deal with the chronology of events which spanned large time periods. An example can be seen in the *History of the Peloponnesian War*, by Thucydides. Although he was very careful in describing events on the timescales of hours, days, weeks, and months, he never placed the war itself in a larger context of historical events—in other words, he never gave the exact years of the beginning and end of the war, which in our contemporary context would be the first thing we would think of asking about such a thing. Indeed, Thucydides would refer to 'the first year of the war', 'the second year of the war', and so forth.

One of the most remarkable Greek innovations was the central role they gave to geometry. While the Egyptians and Babylonians developed and used a wide range of geometric techniques, and indeed it is quite likely that they developed most of those the Greeks later took over, they always did so in a purely practical and indeed utilitarian context. Geometry was a tool for building ships or houses, for dividing land, and so on. The Greeks, on the other hand, made it a crucial ingredient in the construction of their world view.

This period can be conveniently divided into two distinct parts. The classical period, starting in the sixth century BCE, was the high point of the development of Greek culture, and in particular of astronomy and physics. Note that this development was not confined to modern-day Greece, but already included parts of the eastern Mediterranean, especially Asia Minor, and the south of Italy. The subsequent period, usually called the Hellenistic period, traditionally starts with the death of Alexander the Great in 323 BCE and ends with the battle of Actium in 31 BCE (which marks the defeat of the last of the great Hellenistic rulers and the complete military and political domination of the eastern Mediterranean by Rome). This second period is characterised by the spread of Classical Greek culture over a significantly wider area, but at this point a gradual cultural decline was already under way.

2.2 Cosmology

The earliest known Greek mythological descriptions, such as the one in Hesiod's *Theogony*, say that in the beginning there was only chaos (which for the Greeks did not have as negative a connotation as it does today, simply meaning nothingness, or the abyss), and from this the Greek gods and the natural world came into being. They describe the world as a solid disk surrounded by an ocean; below it was the Tartarus (where the spirits of all dead humans were supposed to reside forever), while above it was a region of air, and beyond that another one of aether containing the stars. Thales of Miletus, for example, supported this view.

The city of Miletus, near the coast of what is now Turkey, is the origin of the earliest recorded key contributions of the Greeks to scientific knowledge. Around the sixth century BCE, the first Greek philosophers such as Thales himself, Anaximander, and Anaximenes speculated, without invoking mythology, on how the world was created and how it worked. These Milesians were the first to do what we may legitimately call science: it is recognisable as such not only by scientifically literate amateurs, but even by modern working scientists. Interestingly, they were also the first 'scientists' to be known to posterity by their own names.

For the Milesian school the word 'physics' (from the Greek *physis*, meaning 'nature') is used for the first time as a technical term, denoting properties common to material substances. This school had a materialist philosophy, holding that all existing things were produced from a kind of space-filling primordial substance or fundamental principle (the relevant Greek word was *arche*, meaning 'principle', 'origin', 'beginning', or 'source'), developing from it as a result of the action of various natural forces. These forces were typically arranged in opposing pairs, such as separation and combination or rarefaction and condensation. Although this may look superficially primitive, the important point is that this idea is a legitimate physical hypothesis expressed in logical terms. As such, it is very different from the mythological explanations which were then common.

Why did this development occur in Miletus? One possible explanation is that the Milesians were practical men who had a maritime empire to run, with little time for fanciful and abstract myths. The fact that they had a maritime empire is important because it implies that they were extremely well-travelled for the time. Having regular contacts with foreigners, who would more often than not have completely different myths and views of the world, would lead to scepticism in relation to one's own myths. At the cost of oversimplifying one might say that the legacy of Thales and his successors to scientific thought was the discovery of Nature. Indeed, the idea that the natural phenomena we see around us are explicable in terms of matter interacting by natural laws is not a priori obvious, and the alternative view that they are the results of deliberate or arbitrary acts by gods may be more appealing.

An example of this new frame of mind is Thales' theory of earthquakes, which actually draws on a maritime analogy. Just as sea waves would shake a large boat, the Earth—which was presumed flat and floating on a vast ocean—could sometimes shake or even crack as a result of disturbances in this ocean. This is to be contrasted with the common Greek belief at the time, which was that the earthquakes were caused by the anger of Poseidon, the god of the sea. Again this may seem simplistic to us, but the main point here is that the gods were simply left out in analysing and explaining these phenomena. The Milesian view was that Nature was a dynamic entity which evolved in accordance with some laws (even if they were very imperfectly understood ones), and not the result of constant micromanagement by a bunch of gods.

All these Milesian philosophers tackled the question of the origin of the Universe, and in particular what things were ultimately made of. As has already been mentioned, Thales suggested that in the beginning there was only water, so somehow everything was made from that. Anaximander supposed that there was initially an indefinite, boundless chaos, and the Universe grew from this as from a seed. He also proposed the first known geocentric model: the Earth was a flat disk, totally fixed and unmoving. The Sun, the Moon, and the planets were in spheres revolving around the Earth, and were described as physical objects—and indeed so were the stars. Interestingly, he thought that the Sun was the furthest celestial object, with the Moon below it, then the fixed stars, and the planets nearest to us. Incidentally, Aristophanes' *The Clouds* ridicules Anaximander and Anaximenes' theory that thunder and lightning were the violent escape of wind that had been trapped in clouds.

By comparison to Thales and Anaximander, Anaximenes description is one step closer to modernity. In his model, there was originally only air (a word which should not be taken too literally, but rather interpreted as a gaseous substance). Then the liquids and solids that we see around us formed as the result of processes of condensation. Thus there was a very simple and indeed homogeneous initial state which gradually evolved into the heterogeneous world we know through everyday physical processes. Although it was quite simple, it is what we would today call a detailed physical model.

The spherical shape of the Earth was unknown to the Milesian school. Thales and Anaximenes assumed that the Earth was flat and rested on its primordial substance (water and air, respectively). Anaximander was the first to maintain that the Earth was suspended in space. He also claimed that the Earth had the shape of a cylinder, the depth of which was one third of its width. As has already been mentioned, the Sun would be highest in the sky, followed by the Moon and below it the fixed stars and the planets.

Meanwhile, not far from Miletus, Pythagoras (ca. 580–500 BCE) was born on the island of Samos, though he later moved to Croton, a Greek town in southern Italy. He was therefore a rough contemporary of Anaximenes. Pythagoras founded what in modern terms we would call a cult: a religious group following a strict set of rules governing everyday behaviour (including diet—for example, no beans allowed!) and also including several philosophical and religious aspects. The latter included a belief in the divinity of numbers, the immortality of the soul, and reincarnation as different creatures. Although he left no writings, his school lasted more than two centuries and had a notable cultural impact in this period.

The Pythagorean school also included a strong numerological component, which undoubtedly came from Babylon, as did Thales' prediction of a solar eclipse. The difference, in a nutshell, was that the Babylonians simply worried about algorithms, while the Greeks started worrying about models. The Pythagoreans thought that the numbers had a physical existence, and that they were somehow the fundamental building blocks of the Universe. In other words, numbers not only provided the means to describe individual phenomena and the relations of different phenomena to one another, but actually provided the ultimate substance of things, and therefore the ultimate cause of every phenomenon in Nature.

The Pythagoreans also believed in a primeval element in a Milesian sense, usually fire, as argued by, among others, Heraclitus of Ephesus (ca. 535–475 BCE) and Hippasus of Metapontum (ca. 530–450 BCE). A dissenting view was expressed by Anaxagoras of Clazomenae (499–428 BCE), who argued that no natural substance could be more elementary than any other. In that case there had to be an infinite number of substances, and everything actually had a little bit of everything else in it. For example, he argued that the food we ingested contained small amounts of hair, teeth, nails, and so on, and our bodies then extracted and made the appropriate use of each of these.

The Pythagoreans were the first to argue that the Earth is a sphere. The discovery is traditionally assigned to Parmenides of Elea (510–450 BCE), who was also said to have discovered that the Moon is illuminated by the Sun and that the morning and evening stars, which the Greeks referred to as Phosphorus and Hesperus, are one and the same planet (something that Homer, for example, clearly hadn't realised). Historians of science debate whether the proposal about the sphericity of the Earth was made on purely aesthetic grounds (since the sphere was seen as the most perfect solid) or also included physical arguments. The Pythagoreans also suggested that the planetary spheres, which we will come to shortly, were governed by common harmonic ratios—the harmony of the spheres that Kepler would also attempt much later. This analysis was clearly aesthetic, not just utilitarian.

A generation later, Empedocles of Acragas (now Agrigentum, 484–424 BCE) and Anaxagoras could provide an explanation for solar eclipses which was at least qualitatively correct: the obscuration of the Sun's fire by the intervening Moon. The arguments they used for a spherical Earth did not survive to the present day; they may or may not have been the same given later by Aristotle, namely that the Moon's monthly phases have boundaries of all types but the Earth's shadow cast on the Moon during eclipses is always circular. Additionally, Aristotle also pointed out that one sees different stars as one travels north or south (and the same stars at different altitudes in the sky), which showed that the Earth was in fact a sphere of no great size. A third piece of evidence (of a somewhat different nature, as it relies on his overall concept of the world) was that, if the Earth is made of earth, it will try to settle as close to the centre of the Universe as possible, and this will produce the overall shape of a sphere. Aristotle also claimed that if you look at the phases of the Moon, you can figure out that it too is a sphere. So spheres were the fundamental building blocks of the Universe.

In the sixth century BCE the main philosophical subject of debate was the nature of the primordial substance, but in the fifth century BCE, the focus shifted to what is known as the problem of change. When we look around us we see things changing all the time, but are the fundamental ingredients of the world changeable? If they are changeable, how can they be explained by reference to something more fundamental? In other words, how does one reconcile this with the feeling that the Universe must have some constant, eternal qualities? On the other hand, if the ultimate reality were unchanging, how could one account for (or indeed even accept) the reality of change?

Heraclitus claimed that 'everything flows', and even objects which might superficially seem to be static had some inner tension or dynamism. Thus perpetual change was the fundamental law governing all things—a view summarised in his famous aphorism 'You cannot step twice into the same river'. On the other hand he also stated that, although individual things were subject to change and decay, the world itself was eternal. In his view, this constant change was the outcome of a perpetual strife of opposites: hot and cold, wet and dry, and so on. This was a view which stemmed from the earlier Milesian one, and which was no doubt suggested by the cycle of the seasons. Heraclitus was still a monist, the primordial element in his view being fire, or more specifically the idea of flux. The viewpoint of Parmenides, on the other hand, was almost the exact opposite of that of Heraclitus: for him nothing ever changed, and any apparent change must necessarily be an illusion, a result of our limited perception of the world.

Notice that a view such as that of Heraclitus immediately raises an important question: how can one study the Universe and obtain certain knowledge about it if it is constantly changing? A possible resolution of the problem of change was provided by Empedocles around 450 BCE. His explanation was that everything was made up of four elements: earth, water, air, and fire. These elements were eternal and unchanging, and ensured a degree of immutability in an otherwise changing world. Different substances were made up of appropriate combinations of these elements, in different proportions.

The term air was to be understood as all gases, water as all liquids (including metals, which could be liquefied by heating them), and so on. The aforementioned pairs of forces of attraction and repulsion acted between these elements and could cause their coming together and separation, and thus apparent change in substances. When cooking, for example, fire and water induce changes in the nature of a substance. This belief in four elements (with a fifth one added for the celestial bodies) lasted for about 2000 years, although it was never universally accepted.

The Milesians never considered the question of general causes. Empedocles was the first to distinguish matter from force, and processes that are the result of two opposite forces, which he called 'love' and 'conflict', which mix together and separate the four elements. Empedocles thus postulated the reality of causes in the physical world, and identified them with forces, which today we would call attraction and repulsion. While most early philosophers were monists (believing that all physical substances ultimately originate from a single source), Empedocles was the first pluralist.

The problem of change is a key catalyst for the development of science, because in trying to analyse what changes (and what does not) in the physical world one is led to new ideas such as that of elements, atoms, and conservation laws, which are all cornerstones of modern physics and chemistry. Indeed, this philosophical process culminated with the notion of atomism, which was developed by Leucippus of Miletus (fl. 440–430 BCE) and Democritus of Abdera (fl. 420–410 BCE). Their original notion was that the world is made of tiny atoms, moving randomly in an otherwise infinite void. These were solid corpuscles that were too small to be seen, and came in an infinitude of shapes. Thus atoms were indestructible and eternal (in agreement with Parmenides), but things were constantly being created and destroyed by the coming together and moving apart of atoms (in agreement with Heraclitus).

Although it is easy to dismiss the four-element theory as hopelessly obsolete, one should note that it makes a lot more sense if we think of it in terms of four states of matter—solid, liquid, gas, and plasma. Indeed, in the cosmology of Empedocles the four basic constituents of the Universe are more accurately called 'roots' (in Greek, *rizdmata*), which can be combined in integer proportions to form compounds. Plato later used the term *stoicheia* or 'elements', which originally meant 'letters', but it should be kept in mind that he was a critic of this view. To give a specific example, bone would consist of two parts earth, four parts fire, and two parts water, while blood was made of equal portions of the four roots. The roots always existed in their own right, but they did not always appear to us because they were sometimes mixed with each other. Interestingly, Empedocles compares Nature to painters: just as a painter, with a few colours, can represent diverse forms of very different things, so Nature, with just a few elements, can create all natural substances.

The importance of numbers for the Pythagoreans led to their emphasis on the role of time in their idea of the cosmos. Philolaus (ca. 470–385 BC, and of unknown origin), is believed to have been the first to introduce the idea of the motion of the Earth—in fact Copernicus cites him as an authority on the motion of the Earth. The Pythagoreans displaced the Earth from the centre of the Universe in favour of a 'central fire'. They believed that the central fire was surrounded by nine bodies (since 10 was a sacred Pythagorean number): besides the classical seven planets, they counted the Earth and also a counter-Earth, called Antichthon. From the central fire outwards the order was: Antichthon, Earth, Moon, Sun, the five classical planets, and finally the fixed stars. A cartoon view can be seen in Fig. 2.1

The Sun and the Moon did not shine on their own but would reflect light from this central fire. The Earth would be one of the planets, creating night and day as it travelled in a circle about the centre, both revolving around the central fire and rotating on its own axis as it did so. This double motion was necessary in order to ensure that the central fire should always remain invisible to the inhabitants of the known parts of the Earth. For a modern analogy, you just have to realise that the Earth will be invisible for someone living on the far side of the Moon. The counter-Earth would lie opposite us, possibly in the same orbit around the central fire, and therefore it wouldn't be visible from Earth either, since we always face away from it. The empirically observed fact

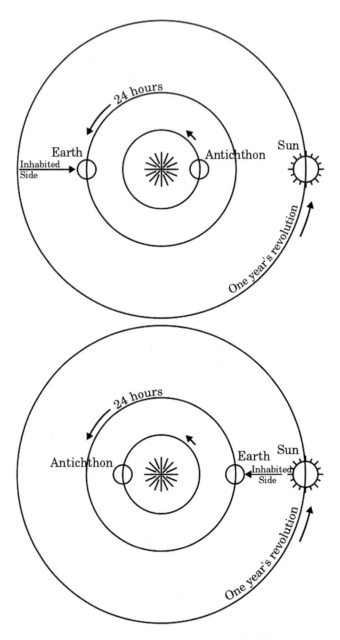

Fig. 2.1 Philolaus' model of the Universe, showing the counter-Earth (Antichthon) orbiting the central fire. The upper illustration depicts the Earth at night and the lower one the Earth during the day. Note that neither Antichthon nor the central fire are visible from Earth (Public domain image)

that lunar eclipses are more frequent then solar eclipses (at a given location, that is) would then be explained in this framework by arguing that the counter-Earth (or indeed more than one of these, in some later developments), and not just the Earth, could block the Moon. The occultation of bright stars by the Moon provides easy evidence for the fact that the stars must be further away than the Moon itself. For example, Aristotle refers to seeing the Moon occult Mars at first quarter.

Outside the fixed stars was an outer fire, then infinite space (*to apeiron*, a term first employed in a physical context by Anaximander), or the infinite air from which the world would draw its breath. The Pythagorean model accounted for the apparent rotation of the heavens, for the day/night cycle, seasons, and motions of the other planets to the same extent as geocentric models. The distances of the Earth and Antichthon to the central fire were assumed to be small in comparison with the distances of the Moon and planets. The Pythagoreans had their numerical mysticism, but they also sought causal laws of Nature, and carried out experimentation. Today we still believe in the power of mathematics and the validity of formulas and calculations, and this is a Pythagorean legacy.

Notice that while this is undoubtedly a physical model, we also see that it is being tweaked to suit philosophical prejudices, while care is also being taken to satisfy observational constraints. The introduction of a central fire must come with a physical explanation for why we don't see it directly, and similarly the introduction of a counter-Earth (to make the desired number of ten objects in the model) also requires a justification for why it can't be seen.

In the cosmology of Plato (427–347 BC) space exists in its own right, as an eternal and absolute frame for the visible order of things—a notion that Newton would later draw on. On the other hand, time is simply an emerging feature of that order. More specifically, time was described as being produced by the revolution of the celestial spheres. The Universe was fashioned by a divine artificer (in effect, the principle of reason) which he called the demiurge. This was neither omnipotent nor the only eternal being (there were also forms and matter which are coeval with it)—and it was certainly not a personal god in the sense of later monotheistic religions. The demiurge created the world by imposing form, law, and order on existing matter, which was originally in a state of chaos. This pattern of law was provided by an abstract and ideal realm of geometrical forms, which were eternal and provided a kind of blueprint for this order.

One can reasonably argue that it was Plato's insistence, at the Academy, on the systematic application of mathematical methods to astronomy, some undoubtedly coming from Babylon and Egypt, combined with the Greek emphasis on geometry, which made astronomy into a science. Incidentally, in Plato's day the order of the planets in increasing distance from Earth was generally believed to be first the Moon, then the Sun, Venus, Mercury, Mars, Jupiter, and finally Saturn. Later on, during the third and second centuries BCE, Greek astronomers gradually adopted a new order based on the periodicity of their observed movements, yielding the Moon, Mercury, Venus, the Sun, Mars, Jupiter, and Saturn.

Plato's *Timaeus* contains his views on the physical world, which formed the basis of modern cosmology for the first 12 centuries of the common era—as will be discussed later, this was only superseded by the rediscovery of Aristotle in the twelfth century. The 'forming artificer' made the world and gave it the most perfect shape, that of a sphere, and provided it with a single uniform motion. The distances of the planets were in the following proportions: Moon 1, Sun 2, Venus 3, Mercury 4, Mars 8, Jupiter 9, Saturn 27. It's clear from the *Timaeus* that Plato assumed that all the planets moved in the same plane, and that the heavenly sphere of fixed stars moved around the Earth in 24 h. Plato's *Phaedrus* also described the Universe as a sphere, and divided it into two main regions: the supra-celestial space was occupied by eternal ideas, while the infra-celestial space was the transient region of sense and appearances.

Plato's god was not the most important thing in the Universe (instead, the forms were), nor was it the only god (he had many assistants). Moreover he was not omnipotent (he had to cooperate with various natural forces), and did not create the Universe from scratch but used the materials already at hand. People were not created by god (but by his junior assistants). He was also an impersonal god, a master craftsman. In a nutshell, he was a Pythagorean god—the principle of reason. He had to be content with the fixed laws of Nature and the intrinsic properties of things.

Plato's cosmos was based on the notion of the perfection of the spherical shape. A spherical Earth was enclosed in (or, perhaps more accurately, at the bottom of) the spherical envelope of the heavens. He assumed various circles on a plane of the celestial sphere, marking the paths of the Sun, Moon, and planets. He understood that the Sun moves around the celestial sphere once a year on a circle (which today we call the ecliptic) that is tilted in relation to the celestial equator. The Moon makes a monthly circuit of approximately the same path, and the five other planets do the same, each moving at a different rate. A key difference when it comes to the motions of the planets is that they can have occasional reversals. Another one is that Mercury and Venus never stray far from the Sun, while Mars, Jupiter, and Saturn do, and can be found at any angular distance from it. Most importantly, he seems to have understood that these irregularities of planetary motion could be explained by a combination of uniform circular motions. Plato's cosmos was animated, permeated by rationality, and full of purpose and design.

It's important to realise that, while the astronomy of the Platonic tradition certainly aimed to have a mathematical description of the apparent motions of the celestial motions, it did not attempt any explanation of physical reality. This mathematical representation was what Plato himself referred to as 'saving the phenomena'—in other words, finding a way of reproducing the observed motions. Since the heavenly bodies were of divine nature (and obeyed laws that were different from the ones we experience on Earth), there was no connection between the two and that made it impossible for us to know anything about the true physics of the heavens.

Plato tried to uphold the Pythagorean teachings about the numerical harmony which imposed form on matter, but his emphasis on deduction over experiment led to the decline of the Pythagorean school. So mystic elements prevailed over scientific aspects, which is eerily reminiscent of the pseudosciences today.

Plato's view that time was a result of the motions within the Universe was criticised by Aristoteles of Stagira (384–322 BCE). He rejected the idea that time could be identified with any form of motion or change by pointing out that the relation between time and change is unavoidably reciprocal: if there is nothing changing, time cannot be measured, and without time, change cannot possibly occur. He did agree that, since circular motion was uniform and eternal, it was the only type of motion suitable for heavenly bodies and hence it was the perfect way to measure time.

Aristotle's natural philosophy was based on the idea of the permanence of the cosmos. He rejected all descriptions based on notions of evolution, and emphasised that to the extent that any change occurs at all it must be cyclical. This idea was then pushed much further in late antiquity by another philosophical school, the Stoics. For example, Zeno of Citium (335–263 BCE) believed that, when the heavenly bodies returned, at fixed intervals of time, to the same relative positions as they had at the beginning of the world, everything would be restored just as it was before and the entire cycle would be renewed in every single detail—just like a movie that is being played and replayed endlessly. Aristotle explicitly denied the possibility of a beginning, and insisted that the Universe must be eternal. This position of course proved troublesome for his medieval commentators.

Greek science was mainly static. Apart from the motions of the stars, not a single dynamical phenomenon was put into mathematical form by the Greeks. In Aristotle, the discussions of dynamics and laws of motion are qualitative, although he was aware of the connection between time and motion. The aether was invented (probably in the fifth century BCE) in order to maintain the non-existence of a vacuum, and raised by Aristotle to the status of a fifth element in additon to the four postulated by Empedocles. Another philosophical school of note was the Epicurists, who built upon Leucippus and Democritus' theory of the atomic structure of matter and the infinite extensions of the vacuum in the space between and outside the atoms.

Nevertheless, the cornerstone of Greek celestial mechanics was an essentially dynamical principle: the idea that the circular movement of celestial bodies was the only movement that could go on forever (which was another reason for its perfection). The fact that the planetary orbits in our Solar System are circular (to a very reasonable degree of approximation) facilitated the gradual development of ever improved geometric models. On the other hand, the philosophical preference for circular motions also proved to be a major barrier to progress.

2.3   The Size of the Earth

From the fifth century BCE (say, from the age of Plato), and in all subsequent epochs, anyone with any pretensions to an education knew that the Earth was spherical. Only a few Epicureans disagreed with this notion, and Aristotle proved it quite comprehensively. Indeed, Erathostenes of Cyrene (ca. 276–195 BCE) provided a measurement of the actual size of the globe that turned out to be quite accurate.

To name just a few of those who wrote in favour of the sphericity of the Earth, one may mention Isidore of Seville (ca. 560–636), the Venerable Bede (673–735), Roger Bacon (ca. 1220–1292), Sacrobosco (ca. 1195–ca. 1256), Jean Buridan (ca. 1290–1358), and Nicole Oresme (ca. 1320–1382). The latter also discussed in some detail the possibility of the rotation of the Earth. The issue of the origin of the flat Earth myth—which emerged comparatively recently, in the nineteenth century—is discussed in Jeffrey Burton Russell's *Inventing the Flat Earth*.

Although the story is well known, Erathostenes' result stems from a clever use of common knowledge, not from custom-made measurements, so it can't really be considered a scientific experiment. It exploited the fact that Syene (modern Asawn) is in the Tropic of Cancer, so the Sun is directly overhead at noon at the summer solstice. Alexandria is almost directly north of Aswan.

Erathostenes heard from travellers that at noon on 21 June the Sun cast no shadow on a well in Syene, meaning that it was directly overhead. He knew that the Sun always casts a shadow in Alexandria. On 21 June he measured the shadow of an obelisk in Alexandria, and through simple geometry calculated that the Sun was (1/50) of 360° away from the vertical, so the Earth's circumference was 50 times the distance between Syene and Alexandria. Travellers told him that camels could travel 100 stadia a day and needed 50 days to travel from Syene to Alexandria, so the distance would be 5000 stadia and the Earth's circumference 250,000 stadia. We don't know exactly what the length of a stadion is (and moreover there were Olympic and Egyptian stadia, each with slightly different lengths), but we think he got a figure of about 42,000 km, within less than 10% of the current value.

Remarkably, further geometrical reasoning also allows the determination of the size of and distance to the Moon. Based on Aristarchus' calculations and his statement that the distance to the Sun was much greater than the distance to the Moon, Eratosthenes noted that the diameter of the shadow of the Earth at the Moon' distance must therefore be very nearly equal to the diameter of the Earth itself, which he had just determined. By comparing the time for the shadow to completely obscure the Moon during an eclipse to the time from the beginning of the eclipse to the emergence of the first portion of the Moon from the shadow, he found that the diameter of the Earth was about 4 times that of the Moon.

The Sun and Moon have an apparent angular size on the sky size of half a degree, so they have diameters which are 1/120 of their distances. Since the size of the Moon must be about 1/4 of the size of the Earth, it followed that the distance to the Moon must be about 60 times the radius of the Earth, which is again within a few percent of modern values. For an Earth radius of about 6400 km, it immediately follows that the radius of the Moon is about 1600 km, and the distance to the Moon is about 384,000 km.

A second method of measuring the Earth's circumference was used by the Stoic Posidonius (ca. 135–151 BCE), a teacher of Cicero who discovered the dependence of the tides on the movement of the Moon. (This is described by Cleomedes, a Stoic who probably lived at the beginning of the present era.) His insight was that Rhodes and Alexandria are on the same meridian and are separated by 5000 stadia, and that the star Canopus (not seen at all in Greece) was barely visible at Rhodes. When viewed toward the southern horizon there, it stood at 7.5° (1/48 of the circle) above the horizon when due south of Alexandria. The Earth's circumference had therefore to be 48 times 5000 stadia, or 240,000 stadia. With the caveat of the uncertainty in the value of the stadion, this was once again accurate to within a few percent.

However, later writers state that Posidonius' estimate was of only 180,000 stadia. The reason could simply have been that this estimate is closer to Ptolemy's own estimates and was picked up in his *Geography*. Ptolemy's estimate was also based on comparing how high some bright stars would be above the southern horizon in various places, and translating this into a difference in latitude. He rejected Erathostenes' calculation (following Strabo) and got a result that was about 30% too short—to be specific, about 28,960 km. Note that this was due to inaccurate data, not to bad reasoning.

Ptolemy made the further mistake of extending Asia to the east, far beyond its true dimension (180 rather than 130°). In Ptolemy's maps this had the effect of drastically reducing the extension of the unknown world between the eastern tip of Asia and the western tip of Europe. Later calculations of the number of degrees of longitude for Asia were even higher. The philosopher and theologian Pierre d'Ailly (1350–1420) suggested 225° of land and Marco Polo 253, the correct number being 131.

This is historically important because Erathostenes' result was almost lost, while Ptolemy's was the most influential. His result would make the west of Europe much closer to the east of Asia, and Columbus relied on Ptolemy (and also Marco Polo) to argue for the advantages of travelling westwards to Asia when he approached the Portuguese court to seek funding for his trip. Columbus also made a further mistake, underestimating the size of 1° by 10%. The straight-line distance between the Canary Islands and Japan is 10,600 nautical miles, not 2400 as estimated by Columbus in his proposal to King João II. To put these distances into context that same year (1484) Diogo Cão had reached the Congo river, already 5000 nautical miles from Lisbon. Astronomers in the Portuguese court were aware of (and trusted) Eratosthenes' estimate, and thus advised against funding a very long westward voyage when an eastern route was soon to become available. Asia was indeed quite far away westwards, but what nobody knew was that a whole new continent was much nearer.

2.4   Astronomy

We have already mentioned the need felt by Greek mathematicians to 'save the phenomena' and try to explain the apparent irregularities in the motions of celestial objects—especially those of the planets—by showing how they might result from a system of regular circular motions. A second motivation for the development of Greek mathematical astronomy in the last decades of the fifth century BCE was the need to have a better calendar.

Several types of motion can be identified in celestial objects by basic observation. First, all objects have a diurnal motion from east to west. Second, there's an annual motion, as a result of which each particular star rises about

4 min earlier from night to night—this is the difference between the solar and sidereal days. And third, planets move only within the zodiac but have specific proper motions (each with its own speed): usually this is from west to east (rising later each day, and known as direct motion) but occasionally in the opposite direction, which astronomers refer to as retrograde motion. Moreover, planets also oscillate in the north–south direction within the zodiac.

For Anaxagoras, the Pythagoreans, Plato, and Aristotle, the order of the celestial objects, from the Earth outwards was the Moon, the Sun, Venus, Mercury, Mars, Jupiter, and Saturn. Later the Stoics changed it to the Moon, Mercury, Venus, Sun, Mars, Jupiter, and Saturn, and this was used by Ptolemy. Indeed, up to the time of Copernicus, it was almost universally accepted.

In the time of Hesiod, a year consisting of 12 months of 30 days was in common use. Note that there was no such thing as a single Greek calendar—almost every Greek community (or city-state) had its own calendar. These certainly had common elements, but they differed from one another in particular aspects, such as the names of the months and the date of the New Year. As far as is known, all were, at least originally, lunar. The months were typically named after important festivals held in each of them (or sometimes the deities honoured in these festivals). It is also interesting to note that the Greeks had no week: the months were divided into 3 decades of 10 days each—except that in months of 29 days the last decade would have only 9 days. This was therefore akin to the Eyptian decans.

The Athenian calendar is the best known. The year began, in theory, with the appearance of the first new Moon after the summer solstice, and when a thirteenth month was intercalated (again, to keep the calendar reasonably aligned with the seasons), this was done after the sixth month, which was called Poseidon—the intercalary month was then called Poseidon II. Each month could have 29 or 30 days, and all were named after particular festivals held in the month. From early times the Athenians dated their years not in a numerical sequence but according to the archon or magistrate who held office in that year (more rigorously, there were in fact three different archons each year, each with specific roles). An almost complete list of archons is known from 528 BC to well into imperial Roman times. Sparta did the same thing with their ephors.

Names of the months, intercalation rules, and other details varied from one city-state to another. For example, the months in the Antikythera mechanism—a remarkable mechanical model of the cosmos and computational tool, whose exact function is still being actively investigated—match those in Illyria and Epirus (in Northeast Greece) and Corcyra (Corfu), all Corinthian colonies. It's possible (though less certain than in the Athenian case) that the intercalated month was also the sixth one. Unlike the Athenian calendar, this one starts in the late summer or early autumn (the Corinthian year is known to have begun at the autumn equinox).

Most Greek religious festivals were correlated with the full Moon (and some occurred at the full Moon itself), but since they were also associated with agricultural activities, keeping the calendar aligned with the seasons was mandatory. The natural solution to these requirements is a luni-solar calendar in which the months are determined by the phases of the moon (whence they would alternate between 29 and 30 days), but a thirteenth intercalary month must be periodically inserted, as was done in Babylon. It is said that, in Athens, Solon (ca. 630–560 BCE) introduced, in or around 594 BCE, the rule of adding an intercalary month every other year. However, this rule was too inaccurate to keep the calendar aligned for long, and improvements were gradually introduced.

The first longer cycle was known as the *octaeteris*, and was introduced sometime during the fifth century BCE, possibly by Cleostratus of Tenedos (ca. 520–432 BCE). As the name suggests this consisted of 8 solar years and 99 lunar months, 3 of which were intercalary. As the Greeks were keen to start a new month when the new Moon was first visible, arbitrary adjustments were needed from time to time, which caused considerable confusion— Aristophanes makes the Moon complain about this in *The Clouds*, first performed in 423 BCE. A more accurate intercalation rule was introduced by the Athenian Meton in 432 BCE, as previously mentioned. This was a 19 year solar cycle of 235 months, of which 125 have 30 days and 110 have 29 (these are referred to as full and hollow months, respectively), adding up to 6940 days; 7 of the 19 years have an intercalary month. Meton and his disciple Euctemon are considered the founders if the Greek 'school' of astronomy. Later on, Calippus of Cyzicus (ca. 370–300 BCE) improved Meton's cycle with one of 76 years, known as the Calippic cycle, which is a set of four Metonic cycles, one of which has a missing day. Later still, in 143 BCE Hipparchus of Nicaea (ca. 190–ca. 120 BCE) recommended a longer cycle of 304 years. Each of these was slightly more accurate than its predecessors, but this had to be balanced against convenience of use.

A variation of the Metonic cycle is still used today by the various Christian churches to determine the date of Easter. The reason for the different Easter dates in the Catholic and Orthodox churches (which occurs more often than not) is that the same algorithm is applied to different calendars—the Gregorian in the former case and the Julian in the latter.

The earliest geometrical and physical planetary theory was that of Eudoxus of Cnidus (408–347 BCE), a disciple of Plato known only from passing comments by Aristotle and much later Simplicius (VI century). Eudoxus, who

Schema huius præmiffæ diuifionis Sphærarum.

Fig. 2.2 The homocentric celestial spheres model of Eudoxus, as depicted in Peter Apian's 1539 Cosmographia (Public domain image)

is also the first Greek known to have had an observatory, at Cnidus (although very little detail survives as to the observations he made or the instruments he had available therein), introduced the homocentric sphere model, a version of which can be seen in Fig. 2.2. This, as well as its slightly later modification by Calippus, is composed exclusively of uniformly rotating spheres concentric to the Earth, perhaps on the analogy of the sphere of fixed stars, which appears to rotate daily about a central Earth.

In the Eudoxus' model, the Sun and Moon each receive three spheres (one to account for their daily motion, the other two for motion along and across the zodiac), while the planets get four each, since one also needs to explain their retrograde motions. There is also one sphere for the fixed stars, making a grand total of 27. Each sphere rotates around separate poles, at its own particular rate. The aim of this structure was to try to reproduce the various apparent motions of these bodies. For example, the Moon moves in the equator of a sphere that rotates with a period of 27.2 days (thus reproducing the sidereal lunar month); outside this sphere there is another with a period of 18.6 years (thus reproducing the so-called cycle of nodes, which among other things determines the occurrence of eclipses), and outside this one there is a third that moves with a period of 24 h (thus reproducing its east to west journey). In general, the theory works reasonably well, though it clearly fails for Mars, and for Venus it is at least unsatisfactory. However, having planets constantly at the same distance from the Earth means that one cannot explain clearly observable phenomena, such as the fact that planets periodically change their brightness.

The Eudoxus' model implies that the Sun moves with constant apparent velocity among the fixed stars, from which it would necessarily follow that the four seasons should be of equal length—which is contrary to knowledge and experience. In order to fix this and other deficiencies, Eudoxus' disciple Calippus added two spheres each for the Sun (to account for the unequal lengths of the four seasons, discovered by Meton and Euctemon) and for the Moon, and one each for Mars, Venus, and Mercury. Although this was certainly an attempt to better save the phenomena, these models were still purely qualitative, and could not be used for calculating the position of a particular planet on a particular day.

The model had many free parameters, such as the period of rotation of each sphere and the inclination of its rotation axis, but there is no clear evidence that Eudoxus ever attempted to accurately determine these parameters in order to make the model agree with the observed motions of the celestial bodies. This is despite the fact that at this epoch some observational parameters, such as the time required by each planet to return to the same position with respect to the Sun, were known with considerable accuracy—thanks, at least in part, to Babylonian observations, of which the Greeks were well aware. In other words, his model could describe the celestial motions in a qualitatively correct way, but not quantitatively. Eudoxus of Cnidus is also known as the first to propose a solar cycle of 3 years of 365 days followed by one of 366 days, which was introduced by Julius Caesar three centuries later and became known as the Julian calendar.

Aristotelian cosmology is clearly built upon that of Eudoxus. The Universe was seen as a great sphere, divided into upper and lower regions by the sphere of the Moon, which had an intermediate nature. The sub-lunar region was every changing and made up of four elements (fire, air, water, and earth, as already discussed by Empedocles and also Plato). In contrast, the supra-lunar region was eternally unchanging, and made of a different, fifth element (known as quintessence or aether).

Eudoxus had put together his spheres as geometrical pictures that provided a solution to the problem of celestial motions. Calippus' improvements were made with observational motivations, but neither of them was trying to construct a model that would be physically possible: they weren't concerned with the real mechanism of the heavens. Theirs was a purely geometrical construction—in other words, only a cartoon model. Aristotle wanted to do better and have a more realistic physical model, at least to the extent that it would be compatible with his own theory of motion (to which we will come shortly). Aristotle turned the cartoon spheres into tangible physical (material) concentric spheres joined together. This enabled one to build an actual mechanical reproduction of the cosmos (again the later Antikythera mechanism comes to mind), and this paradigm became entrenched in the scientific mentality of the time, and subsequently lasted for centuries.

Aristotle noted that since adjoining spheres must be somehow mechanically connected and since he thought (on metaphysical grounds) that the spheres could in principle have disturbing effects on one another, having a viable model required that the individual motion of each planet must not be transmitted to the others. Hence, one also needed 'neutralising' spheres in between the planets, turning in the opposite direction to the 'working' spheres. Aristotle introduced counter-spheres numbering one less than the number of spheres for each planet. In this way, each planet moved as if the spheres of the planets above it did not exist. The only exception was the Moon, which did not require any counter-spheres since there was nothing below it that might be disturbed by its motions. Going through the whole structure one now finds that there was a total of 55 planetary spheres (34 due to Calippus plus 21 counter-spheres), one of which was a sphere of the fixed stars.

The Aristotelian Earth was manifestly spherical. Aristotle provided one conceptual argument for this (earth is the heaviest element), but also several empirical ones, which were already mentioned above. Aristotle's was thus a mechanistic view of a Universe of spherical shells, some of which carried planets. The first sphere of all, the first heaven, engaged in perpetual circular movement (which it transmitted to the lower spheres), but what moved it? Whatever it was had to be unmoved and eternal. Indeed, Aristotle speaks as though each planetary motion had it own prime mover, so that there were 55 of those altogether. Note that, in the Aristotelian concept of inertia, to keep an object in motion you needed to keep pushing on it! Aristotle's view was also anti-atomistic: a vacuum was philosophically absurd for Aristotle, so the Universe had to be a plenum—hence the need for a space-filling fifth element.

Although the above model quickly became the standard cosmological model, it was by no means unique. Heraclides of Pontus (387–312 BCE), a member of Plato's Academy under both Plato and his successor, clearly taught that the Earth rotates on its axis from west to east once every 24 h. This idea came to be fairly widely known, although it was rarely accepted as true; at most one might accept the point that it would explain the daily rising and setting of all the celestial bodies as well as the standard explanation did. Heraclides might also have held that Mercury and Venus moved around the Sun (as if they were its satellites), though this is uncertain since none of his writings survive. In this proposal the Sun still moved around the Earth, but having Mercury and Venus moving around the Sun would explain why they were always seen close to it in the sky. He also called the world a god and attributed divinity to the planets, and he thought the cosmos to be infinite. Perhaps drawing on Pythagorean ideas, he seems to have considered each planet to be a world in its own right, with an Earth-like body and an atmosphere. That said, these possibilities are speculative since all his writings are lost, and are only known from other writers. Chalcidius is the only author to claim that Heraclides considered Mercury and Venus to circle the Sun.

More interestingly, Aristarchus of Samos (310–230 BCE) was the first (and almost the only) Greek astronomer to adopt a fully heliocentric description, thus anticipating Copernicus by eighteen centuries. Although, again, none of his original work survives, there are many indirect references. In particular, Archimedes explicitly mentions Aristarchus' heliocentrism in his work *The Sand-Reckoner*:

> For he supposed that the fixed stars and the Sun are immovable, but that the Earth is carried round the Sun in a circle [...]
>
> Archimedes (ca. 290–212 BCE)

However, the lack of observational evidence for a moving Earth (for example, the lack of a measurement of stellar parallax, which we will further discuss in the next chapter) provided little incentive to accept this as anything more than a speculative possibility. Indeed, how would you demonstrate that the Earth makes an orbit around the Sun, and that it turns on it own axis, if all that you can resort to are naked-eye observations? It's worth bearing in mind that the first direct observations of stellar parallax came only in 1837–1839, by Friedrich Struve (1793–1864) on Vega, by Friedrich Bessel (1784–1846) on 61 Cyg, and by Thomas Henderson (1798–1844) on alpha Centauri.

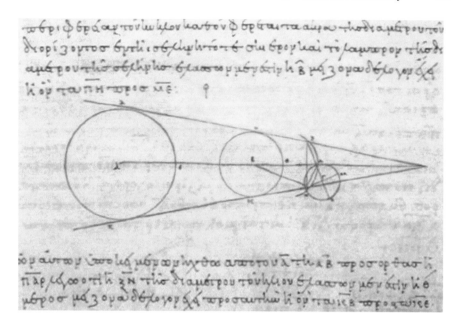

Fig. 2.3 Aristarchus's third century BCE calculations of the relative sizes of (from the left) the Sun, the Earth, and the Moon, from a tenth century Greek copy (Public domain image)

In the model by Aristarchus the Sun was fixed at the centre of the cosmos, while the Earth was considered to circle the Sun as a planet (both rotating on its axis and describing an orbit round the Sun). The sphere of the stars was also taken to be stationary. One of his arguments involved asking why the imperfect Earth should be at the centre of the Universe. Aristotle had argued that the Earth was fixed, among other reasons on the grounds that such a big body could not move. But Aristarchus argued (based on measurements, see Fig. 2.3) that the Sun was bigger than the Earth, so it was even less likely to move. It's also believed that he gave the other planets heliocentric orbits, though the historical evidence does not unambiguously address this point.

The relevant size measurements required observations at first and last quarter. This came to be known as the method of lunar dichotomy, referring to the half-illuminated appearance of the Moon in the first or last quarter. He estimated that the Sun was about 18.9 times as distant from the Earth as the Moon, and had 18.7 times its size—note that observation of total or annular solar eclipses shows that whatever the two numbers are they must be very similar. Now we know that his numbers are underestimated by about a factor of 20, but again this is due to instrumental inaccuracies and not faulty reasoning. Determining the exact moment of first or last quarter is not easy, and neither is measuring very small angles or very large ones (close to 90°). Thus he estimated 87° for an angle that is actually 89.8°, due to the resolution of his instruments and the effects of atmospheric distortions.

It is clear that the model of Aristarchus was a development of Pythagorean cosmology, which had already removed the Earth from the centre of the Universe and put it in motion around the 'central fire'. Aristarchus must have seen the central fire and counter-Earth as superfluous (given that by construction they were necessarily unobservable), and simply removed them. The only astronomer known to have supported these ideas in antiquity was Seleucus of Seleucceia (flourishing around 150 BCE), who is said to have tried to prove the hypothesis. He also suggested that the tides were caused by the Moon (having noted that the periodic variations in the Red Sea tides were correlated with the Moon's position in the zodiac), and even tried to find a physical mechanism to explain it: he thought that the interaction was mediated by Earth's atmosphere.

Aristarchus is the last prominent philosopher or astronomer of the Greek world who aimed to find a physically plausible model of the world. After him, new mathematical developments were still being put to use to describe the observed movements of the planets, but the only real goal of these authors was computational: being able to find the position of each planet at each moment, without any concern regarding the actual physical truth of the system.

2.5   Epicycles

> Had I been present at the Creation, I would have given some useful hints for the better ordering of the universe.
>
> Alfonso X, The Learned (1221–1284)

Even relatively crude observational data, if accumulated over a sufficiently long span of time, can yield relatively precise results. For example, long before the sixth century BCE, the Egyptians were already familiar with the peculiarities of the movements of the planets (including their retrograde motion), and using data available in the second century BCE, Hipparchus was able to determine the average length of the lunar month with an accuracy which is within 1 s of the modern value. By the second century BCE, Greek astronomy had computational planetary theories, although these were still arithmetic rather than geometrical. In other words, there was a separation between mathematical calculations (which heavily relied on Babylonian methods), and philosophical or physical descriptions, which the Greeks built up according to their own philosophical frameworks.

The previously described model of concentric spheres continued to dominate the popular picture of the cosmos throughout the period from Aristotle to Copernicus, but from as early as the second century BCE the Greek astronomers were aware that it was not sufficiently flexible to meet the ever-increasing accuracy of observations. In addition to the problems associated with retrograde motion of the planets, it was clear that the motions of the Sun and planets were not uniform, that the planets had varying brightness (something that can be very easily noticed in the cases of Venus and Mars) and even of size: although opportunities for this were less frequent, it can be seen from the fact that some solar eclipses were total while others were annular. It was equally clear that this could not possibly be explained on the assumption that the Earth was the centre of all these celestial movements and their respective spheres.

In fact, a disciple of Eudoxus, Polemarchos of Cyzicos (fl. ca. 340 BCE) had already realised that the apparent size of the Sun and the Moon are variable, which dealt a death blow to all concentric theories when Autolycos of Pitane (fl. ca. 310 BCE) pointed out that no modification of the system would allow changes in the distance of the planets to the Earth, since all the spheres had the same centre. Moreover, in such a description the retrograde trajectory would always repeat itself in exactly the same way, whereas the observed retrograde motion loops do in fact change from one revolution to the other.

The alternative to spheres within spheres was spheres upon spheres. There were two other equivalent geometrical methods for representing the motions of planets. One was by means of eccentric circles, that is circles whose centre does not necessarily coincide with the Earth. The other was by means of concentric circles surrounding the Earth and carrying epicycles. In either description the Earth was still at the centre of the celestial sphere, but the planets moved around different centres.

The eccentric model preserves uniformity: the planet sweeps out equal angles about the centre of the sphere in equal times, and thus motion seems uniform as seen from there. But seen from Earth the motion appears non-uniform, seeming to slow down as it approaches the aphelion and speed up as it approaches the perihelion. This is therefore adequate to deal with simple non-uniform motions, such as that of the Sun around the ecliptic. In the case of epicycles the planet moves on the epicycle in the same direction that the epicycle moves on the main circle, which is referred to as the deferent. Both can account for non-uniform speed and changing distance, but the latter can also account for the retrograde motion of the planets. Its originator is uncertain—possibly Apollonius of Perga (262–190 BCE) or, even earlier, Heraclides of Pontus—who as you may remember taught that Mercury revolves around the Sun, which in turn revolves round the Earth. It is worth noting that the study of the geometry of conic sections was developed in the late fourth century BCE and peaked early in the second century BCE with a seminal treatise by Apollonius of Perga himself.

In the epicyclic description, the outer planets have deferent periods of one sidereal revolution, while the period of the epicycle is one sidereal year. For the inner planets the deferent period is 1 year, and the period of the epicycle is the heliocentric period: 88 days for Mercury and 225 days for Venus. In other words, for the inner planets the deferent represents the Sun's orbit around the Earth (the equivalent of the Earth's orbit around the Sun), while the epicycle corresponds to the planet's orbit around the Sun. For the outer planets it's the other way around: the deferent accounts for the planet's orbit around the Sun, and the epicycle accounts for the Sun's orbit around the Earth.

In general, the epicyclic system could not give any clues as to the distance of the planets, although it did provide ratios between the radii of the deferent and epicycle (which could be inferred from the length of the retrograde arc). However, it could only do so if one made the further assumption, in agreement with Aristotelian physics, that there was no empty space between the spheres. This notion did not come from Ptolemy, so it must have developed later (possibly in the Islamic world). In any case it was known and universally accepted by the time of Copernicus.

The mathematical relation between the eccentric and epicyclic descriptions was understood by Apollonius, who proved theorems to the effect that an eccentric movement can be replaced by an epicyclic one, in which the centre of the epicycle moves on the deferent with the mean angular motion around the observer, whereas the object moves on the circumference of the epicycle with the same angular velocity but in the opposite direction. The radius of the epicycle is identical with the eccentricity of the eccenter. Similar relations hold for more general cases, and it is therefore a matter of choice (or, in practice, of computational convenience) which description is used.

A complete eccentric circle model for the planetary system is as follows. The Earth is at the centre, the Moon orbits it in 27 days and the Sun in 1 year, probably in concentric circles. Mercury and Venus move in circles whose centres lie along the Sun–Earth line, such that the Earth is always outside these circles. On the other hand Mars, Jupiter, and Saturn move on eccentric circles with centres along the Earth–Sun line, but such that the Earth and Sun are always inside these circles. In fact, the Tycho system, which we will discuss in the following chapter, is one case of this. Although the system was certainly known to and used by Apollonius, nobody claims that he invented it. The Mercury and Venus part had already been found by Heraclides (who made the centres coincide with the Sun), so probably Apollonius' contribution was to provide a more coherent and mathematically robust description. Aristarchus was probably led from this general model to his own model, noting that the Sun orbiting the Earth or the Earth orbiting the Sun leads to the same observable phenomena, if everything else in the model remains unchanged. One may think that the notion of a body moving around a mathematical point must have been philosophically repugnant, so it must have been preceded by motion around a body, but the notion of central motion (due to a cause emanating from the central body) is posterior to that epoch.

Hipparchus of Nicaea developed epicyclic and eccentric models, respectively, for the Moon and Sun, and even derived parameters for them from observation. Nevertheless, it appears that for most computations he still resorted to Babylonian methods rather than to the Greek geometrical models. Hipparchus found the current representations of the planetary motions inaccurate, and collected a significant number of new observations. Computation from geometrical models only became the norm later, with Ptolemy (Claudius Ptolemaeus ca. 90–168).

With the earlier observations and his own, Ptolemy put together a new planetary system. In doing so he further complicated the picture, firstly by using an eccentric circle as deferent, and secondly, instead of making the centre of the epicycle move uniformly in the deferent, by introducing a new point called the equant (see Fig. 2.4), situated at the same distance from the centre of the deferent as the Earth but on the opposite side, and stipulating that the motion of the centre of the epicycle should be such that the apparent motion as seen from the equant should be uniform. Thus uniformity of angular motion was only preserved at that particular point—but that point was neither occupied by anything nor the centre of any motion.

In the epicyclic model, Mercury and Venus have epicycles centred on the line which joins the Earth and the Sun (in other words, their deferents have periods of 1 year, just like the Sun). On the other hand, Mars, Jupiter, and Saturn, which can at times be seen throughout the night (rising in the east at dusk and setting in the west at dawn), have motions described with circles greater than the Sun's, but with epicycles exactly equal to the Sun's cycle, and with the planets at positions in their epicycles which correspond to the Sun's position in its cycle. In principle one can have various levels of epicycles on

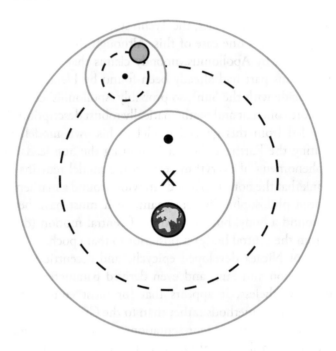

Fig. 2.4 A cartoon sketch of the building blocks of Ptolemaic astronomy: a planet on an epicycle (smaller dashed circle), a deferent (larger dashed circle), the eccentric (cross), and an equant (thick dot) (Public domain images)

epicycles, and if one wants the model to keep pace with gradually improving observations more epicycles gradually have to be added.

An important difference between the Sun and Moon on the one hand and the planets on the other is that the former never have a retrograde motion. The epicyclic description easily accounts for this difference by assuming that the Sun and the Moon move on their epicycles from east to west, while the five planets move on theirs from west to east.

Planetary astronomy seems to have been vigorously pursued in the Hellenistic period, but very little information on this period survives because Ptolemy was so successful in summing up the achievements of his predecessors (and his system was so accurate when compared to earlier ones) that their works dropped out of circulation and disappeared. The only early books that survived were those that were copied, and when you have a better book you throw away the old one. Ptolemy's *Almagest* can be considered the first technical scientific book, only superseded many centuries later by Newton's *Principia*.

Practically all astronomers stuck to geocentric theory, assuming that the stars really did move in circles, and that these did inscribe the Earth, but that their centres did not coincide with the centre of the Earth—in other words, they were eccentric. Note that there is a dichotomy between computational prediction (as carried out by a mathematician) and physical explanation (as proposed by a natural philosopher). Motion is still intrinsically uniform, but the apparent motion as seen from the Earth is not: it appears to move more slowly at apogee (greatest distance from Earth) and faster at perigee. For this to work it's necessary to show that one can choose the line of apses (or, in modern terms, the direction of the semi-major axis of the ellipse) and the eccentricity of the orbit. For example, it was known that superior planets were brightest at opposition.

Nowadays, we know that the planetary orbits are ellipses with the Sun sitting at one focus. Remarkably, with the equant description the position of the equant and the Earth correspond, for a given orbit and to a very good approximation, to the position of the two foci of the ellipse with the same semi-major axis as the size of the corresponding circular orbit. Moreover, given that the angular speed of the planet is uniform as seen form the equant (every 90°, say, of the circle, are swept out by the planet in equal amounts of time), one necessarily has a planet moving faster when it is further away from the equant—and thus closer to the Earth—and more slowly when further away from the Earth, just as in Kepler's laws.

It turns out that in a Keplerian orbit, the planet's angular movement is (by a mathematical accident) almost exactly uniform if instead of seeing it from the Sun you see it from the other focus (that is, the empty focus)—which corresponds to the equant in the Ptolemaic model. Thus if we replace a planet's elliptical orbit by its circumscribed circle, and its Keplerian motion by uniform rotation about the empty focus, we obtain an excellent approximation to its behaviour, with the approximation errors being only to second order in the eccentricity, which is small for the planets in the Solar System.

It's important to realise that although we see a trend of moving further and further from geocentric, uniform circular motion, before Kepler nobody thought of conceiving of planetary motions as ellipses, even though the geometry of conic sections had been developed in Greece as early as the second half of the fourth century BC, culminating in the beginning of the second century BCE with Apollonius of Perga. The circle was an inseparable part of the Greek cosmos. To put it more clearly, the Greeks certainly had all the tools (the mathematical theory of conics, and sufficiently accurate observational data) to have deduced Kepler's laws empirically many centuries before Kepler. The reason they never did was a philosophical prejudice: the circle was key to the Greek world view. Saving the phenomena required adding more and more complexity to the models (compositions of circular motions, eccentrics, epicycles, and equants), but simply abandoning the circles in favour of something that (to our mind, but not theirs) would lead to a much simpler model was a step that was too big for anyone to take.

2.6   Aristotelian Physics

We now make a short digression to discuss Aristotelian physics, both because it is intimately related with its philosophy and cosmology and for the purpose of comparison with more modern notions, whose emergence will be discussed in the following chapters.

According to Aristotle, inanimate matter can have two different kinds of motion, which he called natural and unnatural. (For living beings there is also a third possibility, their deliberate movements, which are dubbed voluntary motion.) Natural motion is the one followed by bodies under the influence of forces arising within the body itself. For example, the downward fall of heavy bodies and the rise of light bodies are respectively due to inner forces called gravity and levity, which furthermore are seen as mutually exclusive—a given body can only have one of them. Unnatural (or violent) motion occurs when something is being pushed. This overcomes the natural motion, and in this case the speed of motion is proportional to the force of the push. The term 'violent' here is meant precisely to indicate that the constant action of an external agent impacts the body and causes the motion.

From a modern perspective it is worth emphasising a key difference at this point: we now view gravity as an external field that causes objects to fall (all of them, incidentally, doing so in the same way), but this is a fairly recent view— even Galileo did not realise that. Before Newton, the falling of a heavy object was considered a natural motion in the further technical sense that it did not require any outside help—or, in modern language, an external force or field.

One difficulty emerging from any attempt to determine the laws of motion empirically in everyday circumstances is that our experimental settings are often far from ideal. In general, the environment includes many sources of friction and resistance, which conceal what would otherwise be a different behaviour. One of the results of these practical difficulties is the pre-Galilean misconception that maintaining a constant velocity requires the action of a constant force. Comparing modern and Aristotelian physics, note that the terminal velocity of an object does depend on the object's mass.

When there are no external obstacles to the motion, natural motion will cause each body to seek its natural place in the Universe, such as a stone falling towards the Earth, or fire rising. Aristotle claimed that the natural state of an

earthly body was to be at rest: it would only move when acted upon by a force. It is also important to realise than in Aristotelian physics bodies could only have one type of motion at a time. All parts of the Earth attempt to fall towards the centre from various directions, and thereby end up making a spherical body.

The Aristotelian physical Universe was finite in extent, thus containing an absolute centre, in relation to which the place and motion of any body could be uniquely defined. Final causes played an essential role in Aristotle's explanation of the physical Universe. The natural motion proper to a body was determined by the proportion of the four basic elements in it. Thus a body composed largely of the element earth would have a greater natural downward speed than one that had a sizeable proportion of the element fire in it. Moreover, a body consisting of earth moved more quickly the closer it was to its natural place (the centre of the Earth and the Universe).

Change and corruption were evident in the world of the Earth, while the heavens were perfect and unchanging, so to the extent that a physical description of the heavens could be given it had to be different. Indeed, the idea that motion could be explained in terms of objects seeking their natural place could not hold for the assumed circular motions of the planets. This made it necessary to postulate that the heavenly bodies were not made up of the four elements but of a fifth one, called aether or quintessence, whose natural motion was circular, eternal, and unchangeable. Circular motion in the heavens was thus also natural in the previously explained sense, and therefore needed no further explanation.

As has already been pointed out the circle was considered a perfect, unchanging, and eternal figure, and therefore not only should the celestial objects move in circles (on the equators of celestial spheres) but they should themselves be spheres. The cosmos was then made of a central spherical Earth surrounded by the Moon, Sun, and stars all moving in circles around it. It may seem strange that all celestial bodies were perfect but nevertheless had to circle the imperfect Earth. However, it is perhaps closer to the Greek frame of mind to say that the Earth was at the bottom of the Universe.

One of the differences between the physics of antiquity and the modern version that most people now learn in school lies in the conception of velocity as a quality of the moving body. In other words, a body could be fast or slow, much in the same way as it could be hot or cold, red or blue. Velocity was the intensity (or degree) of a local motion, and although it was implicitly taken for granted that this intensity had some numerical value, no definition was ever specified, and neither was any attempt ever made to measure it quantitatively. Aristotelian kinematics considered only uniform velocities, simply because it did not possess the mathematical machinery (or a physical context) which would allow a definition of non-uniform velocities.

The need for the action of a force to cause the motion is common to all types of motion. The motion of the spheres would be due to rational souls inhabiting the spheres, which indicates an animistic view of Nature. Any motion that was not natural was violent or forced motion and required the constant action of an external agent. In the sublunar realm, natural motion ceased when an object had reached its natural place, and violent motion ended when the external force no longer existed. In particular, there was no room in Aristotelian physics for any action or force at a distance—which as we will soon see leads to difficulties with projectile motion. The motion of the Coyote in the Road Runner cartoons, horizontally off the cliff and then suddenly falling vertically, is a reasonably accurate visualisation of Aristotle's theory of projectile motion.

For Aristotle there was no concept of a self-sustaining uncaused motion, since an agent was required for all motion. Along the same lines, a true vacuum was impossible because there could be no natural motion there. A medium served both as a cause of motion and as a source of resistance to it. For falling bodies, the acting force was the weight pulling the body down, and the resistance was that of the medium in which the motion is happening (air, water, and so on). Aristotle did realise that a falling object gains speed as it falls, and he explained this as being due to a gain in weight during the motion. If weight determines the speed of fall, then when two different weights are dropped from a high place the heavier one should fall faster and the lighter one more slowly, in proportion to the two weights: for a factor of ten difference in mass, the fall times would differ by a factor of ten.

In discussing natural motion through a medium (say air or water), Aristotle assumed that the speed with which a body moved was directly proportional to the force and inversely proportional to the density of (or the resistance offered by) the medium. Thus for a falling object the natural speed v was directly proportional to its weight W and inversely proportional to the resistance R, so Aristotle's law of motion could be written

$$v = \frac{W}{R}, \tag{2.1}$$

which again exhibits the fact that a heavy body would fall faster than a lighter one, since it would suffer a greater attraction towards Earth. It also shows that a body would move with infinite speed through a void that offered no resistance to motion. Since he considered this impossible he concluded that there could be no void.

That said, one should note two points about the above equation. The first is that writing the equation as such is entirely anachronistic: Aristotle's laws of motion are never expressed in equations in his writings. And the second is that this is actually the equation for the terminal velocity of an object that is simultaneously subject both to a constant force (in this case, its weight) plus a velocity-dependent resistive force.

It may seem surprising, especially if one recalls Aristotle's painstaking observations in other contexts of the natural world (e.g., in his biological work), that he seems to have made no attempt to check these rules in any systematic way. After all, factors of ten in object masses can easily be arranged, and factors of ten in fall times (or a lack thereof) should have been experimentally simple to identify. Obviously, this was not something that he considered important. A possible explanation for this is that the Aristotelian tradition also held that one could discover all the laws governing the Universe by pure thought, without any need for observational or experimental confirmation. Thus, if one is satisfied that a coherent set of laws and explanations has already been achieved there is no incentive for further testing.

Actually, Aristotelian physics was always the most controversial and least robust part of his works, and it did not go unchallenged even in ancient Athens. Thirty years or so after Aristotle's death, Strato pointed out that a stone dropped from a greater height had a greater impact on the ground than the same stone dropped from a smaller height, suggesting that the stone picked up more speed as it fell from the greater height.

Since every motion that was not natural required an external cause and hence an explanation, projectile motion also posed a serious problem. Once an arrow has left a bowstring, it does not fall straight down (as it should according to its natural tendency to seek the centre of the Earth), but instead follows a curved trajectory in its return to the ground. Aristotle met the challenge by assuming that the surrounding air closes in behind the arrow (or any projectile) and pushes it along its curved path. However, this explanation is implausible and counter-intuitive: if you imagine two such arrows passing each other very close to one another and each moving in opposite directions, then the air must also be pushing each of them in opposite directions, but that flatly contradicts the idea that each object (in this case air) can only have one motion at any one time.

The Aristotelian position on projectile motion was always one of the Achilles heels of his physics, and was often attacked. Strato of Lampsacus (ca. 335–ca. 269 BCE), the second of the successors of Aristotle at the Lyceum, argued that a falling body accelerated as it fell, as can be seen from that fact that a stream of water breaks into droplets. (None of his work survives, but it is mentioned by others.) However, an alternative only appeared much later. The so-called antiperistasis theory, introduced by John Philoponus of Caesarea ca. 500. This postulated an inner immaterial force called impetus which was impressed upon a moving object; this would also allow for motion in a vacuum. This idea would be crucial to the development of medieval physics, as we will see in the next chapter.

2.7   From Greece to the Islamic World

An interesting question is why didn't the Greeks, comparatively speaking, advance technology? It has been suggested that this was because they had a slave class and had no need to develop any labour-saving devices, but this isn't a good explanation, as can be seen by comparison to Egypt, who indeed made efforts to optimally exploit great masses of workers (though they were mostly not slaves, but farmers working during the Nile's flood season) in large technical undertakings. The main reason was that Greek mentality was aristocratic, and came to despise manual labour as unworthy of humanity's spiritual destiny. They believed that work without study was worthless, and study for its own sake was the highest level of spiritual activity, superior to the combination of study with any practical purpose.

Their view that the value of science was reduced when it became a means to an end is remarkable in being opposite to the modern one. When applying today for funding for a scientific research project, one is asked to justify any potential technological, economic, or other societal impacts, even if the project itself is entirely conceptual. Clearly, such impacts are highly valued, if nothing else as a long-term return on the initial funding investment. The Greeks would be stunned by our views.

Also related to the separation between science and technology is the Greek reluctance to dissect Nature in the form of experiments. The Ancient Greeks saw no need to improve upon the technical achievements known to them— which were considered to have reached a level at which they could supply the essential necessities of life. Such feelings explain Aristotle's contempt for artisans, craftsmen, architects, and the like. The most priceless possession was pure intellectual curiosity, and the value of a discovery was in no way enhanced by its practical and technical possibilities.

Throughout the many centuries of developments discussed in this chapter, most philosophers were happy to include the heavenly bodies in their pantheon. For Pythagoras they were seen as divine, and Plato expressed his shock at Anaxagoras' atheistic claim that the Sun was a burning mass, and the Moon was a sort of Earth that shone by reflecting sunlight. Anaxagoras was in fact imprisoned for these claims (he was later saved by Pericles, but forced to leave Athens). For Plato the stars were visible manifestations of gods, in which the supreme and eternal being had put light. Plato's view was not a celestial religion for the common folk, but a religion for celestial idealists, and his numerous successors, many of them already working within a Christian framework, made the Platonic view of the heavens highly influential. Even his rival Aristotle defended the divinity of the stars, representing them as eternal substances in perpetually unchanging motions.

Another basic Greek attitude to life was the insistence on moderation when it came to material requirements: everything beyond a 'bare necessity' was considered a superfluous luxury. In agreement with their view of the world as a steady decline from a golden age of gods and heroes, the Ancient Greeks believed fundamentally that the world should be understood, but that there was no need—or possibility—to change it. To some extent this remained the belief of subsequent generations up to the Renaissance. Only then was submission to Nature replaced by a thirst for knowledge as the means for controlling the forces of Nature and harnessing them to man's requirements. This striving for power through knowledge was one of the leading characteristics of the Renaissance. Thus one moves from a passive attitude towards Nature to an aggressive one: one should master it and exploit it for one's own needs.

Without regular new inputs provided by experimentation and technical invention the process of scientific creation became self-limited. Signs of this already appeared in the second century BCE, and were gradually amplified by the penetration of superstitions into the domain of science, especially following the merging of Eastern and Western traditions in the Hellenistic era which led to the growth in popularity of various occult traditions. This decline of creative science became part of the general eclipse of the Ancient World which followed on the disintegration of the Roman Empire and the collapse of political and civil security.

The term 'dark ages' is commonly understood to refer to the period from the fall of the Western Roman empire to the beginning of the Renaissance, but from our point of view this started earlier, with the decline of the Hellenistic civilisation. From the fifth to the twelfth centuries most of the remnants of Classical Greek culture surviving in Western Europe were framed in the context of Neoplatonism (christianised by St. Augustine and the pseudo-

Dionysius), while from the twelfth to the sixteenth centuries Aristotelian philosophy was rediscovered and soon regained a leading role.

The development of new scientific theories virtually stopped, while irrational theories gained ground on science: astrology challenged astronomy, magic increased its role in medicine, and alchemy almost overwhelmed most of the rest of natural science. Although there were some developments in physics, progress is astronomy and cosmology was extremely limited during this period, and it was only re-started with the Copernican revolution.

Meanwhile Islam's Golden Age, which spanned the ninth to thirteenth centuries, brought about key advances in mathematics (including algebra and trigonometry), science (astronomy, optics, and chemistry), and medicine (such as the circulation of blood). At the end of that period, science in the Islamic world also collapsed, and no major invention or discovery has emerged from it for about seven centuries now.

Europe owes a debt of gratitude to Islam for keeping alive much of the classical Greek knowledge for many centuries, and eventually enabling its return to Europe: in their Golden Age there was little of the hostility to science which distinguished Europe during the first half of the Middle Ages. Islamic advances came mainly in practical (as opposed to theoretical) aspects of science. They left cosmological models unchanged: Islamic cosmology drew entirely on the Greek one, and it was simply reprocessed in the early Islamic scholastic centres in Damascus and Baghdad. They made relatively small improvements in planetary theories, but were of course essential in preserving a coherent body of knowledge and information. That said, they also added an important mathematical tool: the 'calculus of chords' due to Ptolemy was developed into the calculus of sines in trigonometry, which is far more convenient and influenced the subsequent development of astronomy.

The origin of the Islamic interest in science can ultimately be traced back to the closure by Justinian of the Neoplatonic Academy of Athens in 529—a distant remnant of Plato's Academy. Scholars from there mainly moved to Persia (roughly modern Iran), and brought with them most of the Greek learning that survived at the time. A scientific institute was later set up in Baghdad after the Muslim conquest of the region. This reached the peak of its reputation during the caliphate of Ma'mun (813–833), who was himself an astronomer. Some of the Greek texts, including Aristotle's *Physics*, went through a process of multiple translations, from the original Greek into Syriac, Arabic, Hebrew, and finally into medieval Latin; the multicultural nature of this effort is highlighted by the fact that many of the scholars who wrote in Arabic were in fact not Arabs but Persians, Jews, and Nestorians.

By the end of the ninth century most of the works that had emerged from Greece had been translated into Arabic, including Ptolemy's great 13-volume astronomical book *Mathematike Syntaxeos Biblia*, sometimes known as the *Megale Syntaxis* (the *Great Treatise*) for short, and also by its Arabic title, the *Al Magest* (in full, *Kitab al-Madjisti*, it was translated into Arabic in 827). Knowledge of Greek science and technology, combined with Iranian and Indian traditions and enhanced by their own scientific studies and inventions, spread across the Islamic world, including Sicily and southern Italy as well as Moorish Spain and Portugal, where by the twelfth century the main centres of learning were Cordoba and Toledo (the *Al Magest* was translated there from Arabic to Latin in 1175). In this way these works re-entered Europe, and their rediscovery stimulated an interest in the natural world that ultimately led to the Renaissance.

The Islamic calendar is one of the few remaining purely lunar calendars; it contains 12 lunar months, and is therefore shorter than the solar year by slightly more than 11 days. The new crescent Moon, which marks the beginning of the lunar months, had to be watched out for and established by two 'witnesses of the instant'. An interesting aspect of the Islamic world was its reliance on mathematically educated persons who could determine the astronomically defined times of prayer and the direction of Mecca—the latter is a non-trivial task on a spherical Earth. This need led to the development of may portable instruments for the determination of time. The most important of these, which was extensively used not only by Arabic but also by Latin astronomers, was the astrolabe (see Fig. 2.5). This instrument was a Hellenistic invention: it was known to and popularised by Ptolemy in the second century of the common era, but the underlying mathematical theory can be traced back to Ptolemy's great predecessor Hipparchus, in the second century BCE. The standard form of astrolabe used in medieval Europe, however, was derived from the Muslim type found in southern Spain and Portugal.

The astrolabe was an early form of analogue computer—the astronomical equivalent of an abacus. Its primary function was to allow the user to solve problems of spherical trigonometry in an approximate but computationally efficient way, which shortened the required astronomical calculations. An astrolabe was always designed for the latitude of a particular place. In broad terms it worked by projecting the celestial sphere onto a plane, and the computationally convenient ways of doing this projection depend on latitude. Its most important use was to determine the precise time of day or night from an observation of the altitude of the Sun or one of the bright stars, a catalogue of which was included on the instrument itself. A typical accuracy for such a calculation might be around 15–20 min, which may sound rather crude by

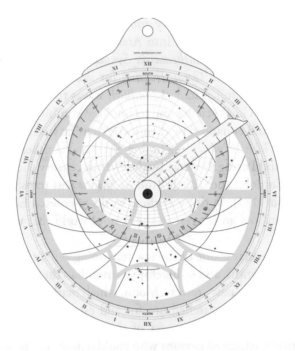

Fig. 2.5 A computer generated planispheric astrolabe for a latitude of 45°, and an astrolabe made of gilded brass from about 1540–1570 (Public domain images)

our modern time-conscious standards, but is nevertheless sufficient for most purposes, especially considering the relative simplicity of the calculation—which could be done in a few minutes, for anyone trained in the use of the astrolabe. The various scales engraved on it also enabled the determination of the positions of the fixed stars in relation to the horizon, and of the Sun, Moon, and planets in relation to the stars.

3

From the Renaissance to Modern Science

We start in medieval times, briefly considering the prevailing world view and emphasising their version of Aristotelian physics. We then look at the motivations and causes for the Renaissance, focusing on the rebirth of astronomy. The remainder of the chapter discusses in more detail the contributions of four seminal figures on the path to modern science: Copernicus, Tycho, Kepler, and Galileo.

3.1   The Medieval World View

In a certain sense—and perhaps with a slight exaggeration—one may say that what characterised the Middle Ages, as compared to Classical Greece or our modern society, was a lack of precision. This is both a reflection of the lack of the necessary tools, and of their own different priorities. The vast majority of people, including many of the most learned, were far less concerned with the passage of time in their daily lives than we are today.

For example, it was difficult to measure time. Water clocks were rare and costly, while sundials were of limited use in regions of Europe were the sky is often cloudy. This general difficulty can be seen in many surviving documents from the epoch, in the lack of precision with which many events—even important ones—were recorded, and specifically in the way people dated their letters. Until the fourteenth century the only part of society with a specific interest in measuring time to a high level of accuracy was the Church, in order to properly determine the canonical prayer times. They provided the

© Springer Nature Switzerland AG 2020
C. Martins, *The Universe Today*, Astronomers' Universe,
https://doi.org/10.1007/978-3-030-49632-6_3

motivation for the introduction of the mechanical clock in the fourteenth century.

It is thought that as late as the fifteenth century few people knew the current year of the Christian era, for the simple reason that it was not used much in everyday life, and was used differently in different places. For example, there were various different conventions for the beginning of the calendar year—the current choice of 1 January was by no means unique. Thus people seldom dated their letters, and when they did need to do it, they often resorted to the year of the king's reign. Gerald Whitrow provides a compelling illustration of this: if you set out from Venice on 1 March 1245 (the first day of the Venetian year), you would be back in 1244 when you reached Florence but already in 1246 if you went to nearby Pisa. You would be back in 1245 when you reached Provence, and if you reached France before Easter (April 16) you would again find yourself in the year 1244. The obvious way out of this mess was to ignore it, and note only the month and the day, on which there was a wider consensus. Letters were frequently dated in this way (using the year at the place of origin if that was considered necessary), but at least in relatively informal contexts a more convenient dating system was to use festivals and saints' days—a tradition that is still not uncommon today. As an aside, the practice of numbering the days of the month, starting from one, came to the West from Syria and Egypt in the second half of the sixth century, but only became widespread around the eleventh century.

The medieval view of the cosmos drew from a variety of ancient, mostly Greek sources. The Siege of Toledo by Alfonso VI of Leon and its subsequent fall in 1085 allowed much easier access to a vast library of classical and Arabic knowledge. For example, Ptolemy's *Almagest* was translated from Greek in Sicily in about 1150 by Hermann of Carinthia (ca. 1100–ca. 1160), and from Arabic in Toledo in 1175 by Gerard of Cremona (ca. 1114–1187). As discussed in the previous chapter the sphericity of the Earth was beyond question, and explicitly mentioned by many writers. Several ancient estimates of its circumference were widely known, though which was the more accurate was a matter of debate. Often the deciding point of the discussion was one of authority. The cosmos described by twelfth century scholars was a simplified version of the Platonic description: homogeneous, and composed of the same elements from top to bottom. Aristotle's quintessence and his strict dichotomy between the celestial and terrestrial regions only became well known in the thirteenth century, and it gradually replaced the Platonic view.

This simplified version of the cosmos ignored most of the astronomical details, and one can say that attempts to save the phenomena were ignored. Each planet had a single sphere to describe its motion. In addition to the planetary spheres, defining the outer limits of the cosmos, there was the sphere of fixed stars (sometimes called the *primum mobile*), though in some more detailed descriptions there were three or four spheres, with the aim of accounting for the daily motion, precession, and a hypothetical motion called trepidation which as it turns out does not exist at all. Therefore the total number of spheres was 10 or 11. The eighth to eleventh spheres also had important theological roles. The biblical firmament was most frequently identified with the eighth sphere of the fixed stars. The outermost sphere was the Empyrean sphere, which, in contrast to the perpetually moving orbs it enclosed, was immobile. Last but not least, it was also conceived as the dwelling place of God and his elect.

Medieval natural philosophers frequently asked what caused these spheres to move. Most believed that each orb was assigned an immaterial, intelligent being (often identified with an angel) whose voluntary action made its assigned sphere move with uniform circular motion. A small minority held that God created the orbs with an intrinsic power for self-motion. Jean Buridan invoked a divinely impressed force, or impetus (which we will discuss further in the next section), that enabled the orbs to move forever through a celestial aether that was assumed to offer no resistance to their motion.

The re-emergence of Aristotle's works produced a dilemma for medieval natural philosophers and theologians. The Bible states that the world was created *ex nihilo* by a divine power, who would eventually destroy it. But Aristotle's *On the Heavens* contains an equally explicit but opposite message, namely that the world was eternal: it could not have had a beginning and could never come to an end. The Aristotelian Universe simply exists, and there is no room for a creator, so God's only possible role in such a Universe was that of the prime mover.

Reactions to Aristotle's works varied. Most medieval scholars accepted the bulk of Aristotelian philosophy and simply thought that he was wrong on this particular point. Some, like St. Bonaventure (1221–1274), were convinced that the temporal creation of the world could be demonstrable, thus refuting Aristotle's claim on this point. By contrast, Thomas Aquinas (ca. 1225–1274) attempted to reconcile the two views, arguing that there was no contradiction in the possibility that God could have willed the existence of the world without a temporal beginning, making it both created and eternal. His basic claim was that the act of creation (and hence the existence of a creator) could not be demonstrated logically, but could only be understood through revelation.

Although all medieval scholars agreed with Aristotle's arguments for the Earth being immobile, several nevertheless thought that it was worth discussing whether the Earth rotated on its axis, and what would happen if it actually did so. The most detailed discussion of the implications of a rotating Earth came from the aforementioned Buridan and Oresme. For different reasons, they both rejected this possibility, but agreed that if one were to assume that the heavens were immobile, then the Earth's daily rotation saved the astronomical phenomena just as well as the traditional explanation. Note that in this discussion there was no thought of removing the Earth from the centre of the cosmos. The discussion was restricted to what we would call, in modern terms, a question of relative motion: a daily rotation of the Earth about its axis could replace a daily rotation of each of the celestial spheres. On the one hand there was an issue of simplicity (replacing many fast rotations by a single slow one), but of course other philosophical and theological arguments also made the two explanations different. And these arguments carried an enormous weight and were the ultimate deciding factor in such discussions.

Oresme actually went further than this. He also discussed whether it was possible that several worlds exist, whether the heaven was of another nature than inferior things, and whether all the stars received their light from the Sun or some stars produced light themselves. Nicolaus Copernicus (1473–1543) found the arguments of Buridan and Oresme in favour of a rotating Earth worthy of inclusion among the supporting arguments for his heliocentric system, in which the Earth was assigned both a daily rotation and an annual motion around the Sun. Medieval, Aristotelian-inspired cosmology remained functional until the end of the seventeenth century, by which time the successive contributions of (among others) Copernicus, Tycho, Kepler, Galileo, and finally Newton led to an entirely new standard cosmological model.

France and England seemingly emerged from the dark ages earlier than the rest of Europe, but they would be dramatically slowed down by the so-called Hundred Years War (which was neither a single war nor exactly one hundred years long). The vanguard of the revival thus moved eastwards, into Germany and especially into Italy. In the fifteenth century the sources of knowledge from antiquity, mainly preserved in the Islamic world, began to flow faster as well as more abundantly. From the beginning of the century the Byzantine Church, seeking help against the Turkish threat, gradually came into closer contact with the Roman Church. One of the consequences of these contacts was that knowledge of Greek rapidly increased among Western scholars.

Numerous Greek manuscripts were brought into the West, especially to Italy, where they were collected, studied, and eventually also translated into Latin. Now one could directly read what Aristotle, Ptolemy, and very many others had actually written, many centuries earlier. One could therefore recover the pure unadulterated knowledge of the ancient writers, instead of being exposed to the risks of the corrupt and often obscure translations through Syriac and Arabic. If you are doing a translation you may be less than fluent in one of the two languages, or may simply fail to understand the point that the original writer was trying to make, and any such errors, once introduced, are more likely to be amplified by successive translations or copies than they are to be corrected by a zealous and critically thinking translator. Moreover, the possibility of printing the books (following a careful checking and correction of the text to be printed) put an end to the numerous copying errors in manuscripts: once typeset, all books are exact copies of their typeset source.

Indeed, the printing press was a key catalyst to the Renaissance. The required technology (metallurgy, oil-based ink, manufacture of punches) was mainly developed in the Mainz area, though Chinese printing had been around since the 800s at least. It has been estimated that the number of books printed in the 50 years after Gutenberg approximately equalled the number produced by all the European scribes in the preceding 1000 years. Between 1450 and 1500, printing presses appeared in more than 200 towns in Europe, and the numbers more than doubled in the following century. Moreover, printed books were cheaper (by a factor of at least five, and possibly as much as ten) than the previous hand-written ones, in addition to dramatically reducing the production time. At least a million books (not titles) had been printed in Europe by 1550.

The impact of printing on medieval society cannot be overemphasised: it was at least as great as (if not larger than) that of computers and the internet on modern society. Up to around 1500, knowledge did not change: all that there was to know had already been identified and written down long ago and far away, though some of it may have been lost and needed to be rediscovered. Afterwards, not only could old knowledge change and new knowledge be produced, but these developments could also be disseminated at an unprecedentedly fast pace.

An obvious practical issue in this context was how to retrieve old knowledge that had been temporarily lost, or possibly changed by the gradual propagation of errors. It was therefore essential to go back to the original sources and go carefully through the existing body of knowledge, to check whether it was indeed reliable and in agreement with that of the perceived Greek Golden Age, or whether reforms and changes were in order. It was this preoccupation with going back to the original Greek sources that can be seen more clearly today in the arts of the time: think of the role of Greek myths in the painting, sculpture, music, or literature of the time.

In the fifteenth century, the work of systematically reforming European astronomy was started by Georg Peurbach (1423–1461) and, following his death, continued by his student Johannes Muller, known, using the Latinised version of his place of birth, as Regiomontanus (1436–1476). As in other areas, their focus was on identifying and correcting errors in astronomical texts (and in particular in those of Ptolemy), by systematically going back to his original Greek texts. A secondary aim of this effort was to obtain a deeper insight into the thoughts of the original authors. There was a clear notion that a lot could have been lost or oversimplified throughout the centuries, and therefore that there was a reasonable expectation that previously overlooked information might still be recovered from (as it were) the mouth of the original author. Thus Peurbach and Regiomontanus not only republished the *Almagest* but in doing so also dramatically raised the level of theoretical astronomy in Europe.

The models that mathematical astronomers used at the time to predict planetary positions were descended from those in the *Almagest*, but did not address the issue of what caused planetary motion at all—they were merely a computational algorithm, with no claims of being a physical model. They had reached the West through Arabic sources and were popularised by Peurbach. In his *Theoricae novae planetarum* (thought to have been completed in 1454, but only published in Nuremberg ca. 1472), he consistently employed sets of three-dimensional spheres to describe the basic motions of the planets before considering the two-dimensional models descending from the *Almagest*. Peurbach's book became the standard introductory astronomy textbook for over a century and a half. Peurbach also wrote, with Regiomontanus, the *Epitome of Ptolemy's Almagest* (published in 1496), the clearest and most accurate exposition of Ptolemaic astronomy in the fifteenth and sixteenth centuries.

Astronomy in the fifteenth and sixteenth centuries stood nearer to the centre of practical life and occupied people's attention more than any other science. It provided a technical basis for astrology, whose popularity grew after the Black Death pandemic spread across most of Europe from 1348. Astrology was also part of medicine: the process of diagnosing and treating a patient's ailment including casting a horoscope for the person. Princes and many nobles employed 'mathematicians' whose main role was that of astrological advisers, but humanists such as Pico della Mirandola (1463–1494) were already denouncing astrology as superstition. The model of Aristotle and Ptolemy provided a useful computational algorithm for these purposes: it described the motions of the heavens as they seemed to be for common people, it matched observations to a sufficient degree, and therefore it was a reasonably accurate tool for the necessary forecasts. And, at a more abstract level, it matched the commonly held philosophical opinions about the rest of Nature.

Until the sixteenth century, astronomers were on one side of a deep dichotomy which had physics on the opposite side. University students would encounter astronomy as one part of the *quadrivium*, being taught about the structure of celestial circles, the geometry of planetary mechanisms, and the mathematical methods used to calculate the planetary positions required for casting horoscopes. On the other hand, the Aristotelian physical model for the heavens (to the extent that such a thing existed) was not described in his *De coelo*, but in his *Metaphysica*, and that text was taught by the philosophy professor. Among many other cases, this dichotomy is manifest in the fact that, when Galileo was negotiating his new position at the Florentine court with Cosimo de Medici's emissary, the bottleneck in the negotiations was not his salary, but what his job title would be: he insisted on being Mathematician and Philosopher to the Grand Duke. In other words, he wanted the authority not just to make mathematical astronomical models or hypotheses, but to discuss how the Universe was really made.

3.2   Medieval Physics

Omne quod movetur ab alio movetur.

Aristotle (384–322 BCE)

There is an interesting parallel in the way that Greek and modern science emerged, as a radical departure from what preceded them. Greek scientists (or more accurately, philosophers) replaced *mythos* by *logos*, but the philosophical and theological baggage accumulated over centuries had come to be a heavy a burden, so their seventeenth century successors had to discard the baggage and make science an independent sphere of human thought. And herein is therefore a significant difference between the two revolutions: the first married science to philosophy, while the second divorced the two. Although the term 'natural philosopher' was used to refer to scientists for a rather long time after this period, Galileo and his successors were not philosophers in any reasonable sense of the word. The bridges that today remain between the two fields, interesting though they certainly are, only pertain to epistemological discussion on the meaning of science and its results, but have no impact on its methods.

Peurbach's system had reconciled Ptolemy's model, with its eccentrics and epicycles, with the earlier Aristotelian description involving homocentric spheres. The solution was to connect together the Ptolemaic multi-sphere mechanisms for each planet such that the peripheries of each set of spheres were adjacent to each other (avoiding any intervening vacuum) and homocentric. The Ptolemaic equant was more problematic. On the one hand, if one was sitting at the equant one would see a uniform motion, which was therefore perfectly fine. On the other hand, the physical motion of the epicycle centre around the deferent was not uniform, though in the late-medieval period Islamic astronomers found a possible way to reconcile the two descriptions.

Although the views of Aristotle dominated physics, several problems with it had been identified, as was already pointed out in the previous chapter. In about 533, John Philoponus (ca. 490–ca. 570), precisely in a commentary on Aristotle's *Physics*, stated that taking two masses that greatly differed in weight and releasing them from the same elevation, one would find that the ratio of their fall times didn't follow the ratio of the weights: instead the difference was seen to be small, and if the weights did not greatly differ, then the difference was either non-existent or imperceptibly small. Moreover, he also objected to the view that the medium through which an object moved was a causal factor, in the way Aristotle claimed that it was, and he saw no difficulty with motion in a vacuum.

Instead of the Aristotelian law of motion which we discussed in the previous chapter—recall Eq. (2.1)—he proposed the following alternative:

$$v = W - R. \tag{3.1}$$

Note that in this case there is indeed no mathematical problem with motion in vacuum, where the resistance vanishes. To explain projectile motion, Philoponus argued for the existence of an impressed force. A significant aspect of these theories of impressed force was that they all saw this impetus as an intrinsic property of each object that was the cause, rather than the effect, of motion.

Medieval philosophers gladly adopted the impetus theory of Philoponus. This is known to have been taught at Paris University from about 1320, and it was further developed by Jean Buridan. The mover transmits to the moving body a certain amount of impetus or force, enabling the body to keep moving in the direction in which the mover sent it. Once released, this impetus will gradually weaken due to the resistance of the air and eventually it will vanish completely, at which point natural motion is recovered. This impetus was supposed to be proportional to both the object's weight and its velocity. It is worthy of note that this is formally equivalent to the definition of linear momentum in classical physics. Nevertheless, the two have different interpretations in their respective dynamical theories: here impetus is a force with the same status as gravity, levity, or magnetism, and it is a cause of the motion, while in modern terms momentum is a consequence of the motion.

A group of fourteenth century thinkers known as the Oxford Calculators (and almost all associated with Merton College, Oxford) were the first to take a more systematic and logico-mathematical approach to philosophical problems. This school drew from the earlier tradition of people such as Robert Grosseteste (ca. 1175–1253), and influenced others such as William of Ockham (ca. 1300–1350). There were four main members of this group, all contemporaries: Thomas Bradwardine (ca. 1290–1349), William Heytesbury (ca. 1313–1373), Richard Swineshead (fl. 1340–1360), and John Dumbleton (ca. 1310–1350). Their most important contribution was creating a school of mechanics which distinguished kinematics (describing how a given motion proceeds in space and time) from dynamics (investigation of forces as the causes of motions). The Oxford Calculators emphasised the former, and in particular provided definitions of concepts such as instantaneous velocity and uniformly accelerated motion which were acceptable given their main limiting factor: the lack of a suitable concept of limit, and the corresponding tools of differential calculus, which would only be developed much later by Newton and Leibniz.

Around this time an important conceptual change took place in the context of the discussion regarding impetus. In medieval mechanics violent motion was invariably analysed in terms of cause and effect, the cause being some physical force in Nature, and the effect the motion that was thought to be due to this force. Our modern principle of inertia would have been unacceptable to medieval scholars: the necessity of a cause for the motion could only be expressed as some direct relation between the applied force and the resulting velocity. The motion of projectiles along curved paths was the standard problem in violent motion, and determining the origin of this force became an increasingly pressing problem in the fourteenth century, with the ever more widespread use of artillery in battle: understanding this relation would potentially bring significant military advantages.

The traditional Aristotelian theory assumed that the force acting on the projectile came directly from the air (which has received it from the hand, or the catapult in the case of a projectile). At some point during the trajectory the impetus is exhausted and the projectile falls vertically back to the Earth. Ockham pointed out that, if one considers two arrows flying close to each other in opposite directions and wishes to explain their motions as an effect of the surrounding medium, then the air must be considered to simultaneously have a motive force in two opposite directions, which is absurd. Ockham's view was simply that motion, once it existed, did not require a force to be sustained.

Ockham's student Buridan developed the earlier impetus theory into a more quantitative version in which the mover of an object transmits to it a power proportional to the product of the mass in the object times its speed. This impetus was a permanent impressed force that could in principle remain indefinitely in the body, though in practice it could be removed by another external agent. This does allow for the possibility that a moving body left to itself would continue in motion forever—as in the case of the celestial spheres in uniform circular motion. Buridan's student Oresme studied uniformly accelerated motion and proved what is known as the Merton theorem (which the Oxford Calculators had also studied in some detail), stating that a uniformly accelerated body travels the same distance as a body with uniform speed whose speed is half the final velocity of the accelerated body.

Notice that in the absence of any resistance—in vacuum, for example—the impetus would carry a projectile with constant velocity along a straight line to infinity, which is the kind of motion described by the law of inertia. Buridan also used impetus to explain accelerated free fall motion: gravity initially acts alone, but as the body falls it gradually acquires velocity and therefore impetus, and the effect of the impetus adds to that of gravity and therefore increases the velocity. Later on Domingo de Soto (1494–1560), a Spanish Dominican friar, provided the first accurate statement of the law of free fall, in his textbook on Aristotelian physics.

Buridan also remarked that circular celestial motions could be explained without the need for the constant work of angels, if one simply assumed that at the creation of the world God imparted a certain amount of impetus to each sphere. Since the spheres moved without resistance (as the heavens were by definition perfect and any kind of resistance was not), their impetus was constant and kept them in a state of uniform rotation. Notice that this was an interesting and non-trivial departure from the firm terrestrial/celestial dichotomy: celestial mechanics was included in a system developed to describe terrestrial motions.

In the thirteenth century, universities emerging across Europe attracted the best scholars of the time. Later, humanists transferred the authority from universities to themselves, and by the fifteenth century the intellectual standards of the universities began to decline—in part because their success led to new ones appearing in increasing numbers. The aforementioned Regiomontanus is a good example of this: in 1471 he settled in Nuremburg, where he established a complete astronomical institute, including not only an observatory but also a workshop for instrument making and a printing office, all this being paid for by Bernhard Walter (1436–1504), a wealthy local citizen. This was the first European research institute funded outside a university.

The culmination of this gradual decay emerged later, in the sixteenth to seventeenth centuries, when many of the protagonists of science left the universities and sought direct funding of various wealthy patrons (most frequently kings or princes) either on a private basis or as members of scientific academies. One unintended consequence of this was that scientific research and teaching became separated, and as a result the ordinary university student no longer became exposed to and acquainted with the most recent ideas.

At first, this process happened on a private, individual basis, but it led eventually to the formation of the first scientific academies. The earliest of these was the Accademia dei Lincei, founded in Rome in 1603 by Federico Cesi (1585–1630). Galileo joined in 1611, but the academy declined in the 1630s following Cesi's death and Galileo's condemnation. It was subsequently revived in the nineteenth century and today it is still the official Italian scientific academy. The 1660s saw the creation of two other academies which would play a crucial role in the scientific developments of the following centuries: the Royal Society of London (1660, chartered by Charles II in 1662) and the Académie des Sciences in Paris (1666), and many other followed subsequently.

3.3 Mikolaj Koppernigk

The development of mathematical astronomy and more persistent efforts to compare models to observations (even though the available instruments for those observations were still comparatively rudimentary) gradually revealed that the model of Aristotle and Ptolemy was not only quite complex but also not entirely accurate in predicting astronomical phenomena such as eclipses, oppositions, or conjunctions. Although several other contemporaries alluded to the heliocentric hypothesis of Aristarchus, it was Nicolaus Copernicus (1473–1543, see Fig. 3.1) who studied it in detail and also argued for its efficacy. This he did first in a short publication (the *Comentariolus* or *Little Commentary*, circulated in manuscript form not later than 1514) and then in the book *De revolutionibus orbium coelestium* (*On the revolutions of the celestial orbs*, published in 1543) which he only saw published on his deathbed. A preface, known to have been added by Andreas Osiander (1498–1552), a Lutheran theologian, stated that the new view should be seen as a hypothesis that allowed accurate computations rather than an established physical theory, no doubt hoping to appease both Aristotelians and theologians; however,

Fig. 3.1 Portrait of Copernicus in the Toruń town hall, by an anonymous artist ca. 1580 (Public domain image)

further evidence suggests that Copernicus himself was convinced of the reality of his model.

Copernicus was born on 19 February 1473 in Thorn (now Torun) and died on 24 May 1543 in Frauenburg (now Frombork), both in modern Poland. He was descended from a family of German colonists, and his father had a business trading in copper, and also became a civic leader in Torun and a magistrate. Copernicus (which is the Latinised version of his surname) was the youngest of four siblings. His uncle Lucas Watzenrode, a canon at Frauenburg Cathedral and later Bishop of Ermland, became the guardian of all four following the death of Copernicus' father when he was aged ten.

His education started in Wloclawek and continued at the University of Krakow, where Copernicus learned his astronomy from the *Tractatus de Sphaera* by Johannes de Sacrobosco, written in 1220 but still a canonical text at this time. The astronomy training available at the time would essentially have been a mathematics course describing the model of Aristotle and Ptolemy to the extent that it was necessary to understand the calendar, calculate the dates of holy days, use the stars to navigate at sea, and last but not least calculate horoscopes—which among other uses was considered relevant for medicine.

He then continued his studies in Italy, first in Bologna (where he rented rooms at the house of the astronomy professor Domenico Maria de Novara and effectively became his assistant) and later in Padua, though he eventually took his degree at the University of Ferrara. While in Bologna, Copernicus was informed of his appointment as a canon at Frauenburg Cathedral, a job opportunity provided by his uncle which ensured him a lifetime with a comfortable salary and not too many duties.

The *Comentariolus* (usually called the *Little Commentary*), which he distributed in manuscript form to various friends and other astronomers sometime between 1510 and 1514, contains a first sketch of his theory of a heliocentric Universe. Indeed, it has all the necessary ingredients of his theory. Specifically, it lists his seven axioms, the word being used in the sense that they were the hypotheses that were necessary and sufficient to reach his conclusions. What was lacking in the *Comentariolus* were the mathematical details and a comparison of the model to the observational data, both of which he presumably developed in subsequent years. (So the *Cometariolus* may therefore be seen as a roadmap for his later work.) Note that Copernicus' work was mainly what we would now call mathematical astronomy, and he made a very small number of observations himself, mostly relying on classical data or the data of some of his contemporaries.

The key conceptual points of the Copernican model were that the Sun was near the centre of the Universe (though not exactly at the centre), and not all the celestial spheres had the same centre. The Earth was only the centre of gravity and of the lunar sphere, but it and the other planets moved around the Sun in circular orbits. The distance from the Earth to the stars was much greater than that from the Earth to the Sun, and the apparent motions of the firmament were the result of the Earth's motion: its daily rotation around its fixed poles and its revolution around the Sun were sufficient to explain the apparent motions of the stars, the Sun and the other planets, including the key issue of the periods of retrograde motion of the planets. Epicycles had originally been introduced for the purpose of explaining these retrograde motions within a geocentric model, so explaining them in the heliocentric model was mandatory. Copernicus' answer was that the planets simply move around the Sun at different speeds.

Notice that each individual motion in the model was still circular and uniform, and this was justified using the same philosophical and aesthetic reasons that the Greeks might have used. Copernicus was an Aristotelian in accepting the need for uniform circular motions. One can even go so far as to say that Copernicus swapped the positions of the Earth and the Sun and attempted simplifications, but his geometrical language was entirely

Ptolemaic. In his mind, his aim was probably that of recovering a model that was closer to the original Greek philosophical preferences. It is also striking that the presentation of his results in final form, in his *De Revolutionibus*, closely followed the structure of Ptolemy's *Almagest*. In other words, Copernicus was contributing to the general effort of reforming astronomy by providing a modern version of Ptolemy's classic text. As Kepler later put it:

> Copernicus tried to interpret Ptolemy rather than Nature.
>
> Johannes Kepler (1571–1630)

The Copernican ideas were not new as such—as has been emphasised they were essentially the same as those of Aristarchus, and had been occasionally repeated throughout the centuries by several people—including, of all people, Leonardo da Vinci. That said, Copernicus did provide a unique contribution, because he developed the model's detailed mathematical framework, putting it on a par with what was known for the geocentric model. He further showed its consistency with the existing data, and used it to make further predictions— in other words, he made it a falsifiable scientific theory. Taking the model as a computational algorithm this was a major breakthrough, the practical usefulness of which would be amply demonstrated in subsequent decades. On the other hand, from the perspective of a possible physical model there was one drawback: Copernicus used very different geometrical models to describe the motions of the planets in latitude and in longitude, while for a physically real description one would naturally expect one and the same model to account for both. Whatever physical mechanism runs the heavens there should only be one mechanism, not two.

Copernicus calculated the period of motion of each planet around the Sun, using the data available to him, and also the size of the orbits of each planet relative to the size of the Earth's orbit. It should be noted that this result would not be obtainable from Ptolemy's geocentric model. That model did not have any fixed distance scale, providing only a means to obtain the directions of the planets: only the relative sizes of the various deferents and epicycles were specified. Lengths in Ptolemy's model could be changed at will, provided the proportions between them were unchanged. In the heliocentric model a single movement, that of each planet around the Sun, replaced various epicycles and deferents, and the sizes of the orbits of other planets could thus be compared to that of the Earth. Therefore it was also possible for the first time to talk about a scale for the Universe.

Another contribution by Copernicus was to address three major objections raised by Ptolemy to Aristarchus' model, and all having to do—in modern terms—with the relativity of motion in different frames. Regarding the point that clouds and surface objects on a rotating Earth would fly off, Copernicus answered that the same objection existed (and was even more problematic) in the geocentric model, where distant planets had to move around the Earth at high speeds. To the objection that an object propelled directly upward on a rotating Earth would not fall back to its point of departure but would be left behind, Copernicus answered that such objects, like the clouds and the air, were part of the Earth and hence carried along in its rotation.

Finally, the third and most pertinent point was that of the annual revolution of the Earth around the Sun, which should evince a change in the position of the fixed stars when observed at opposite ends of the Earth's orbit. This effect is known as annual parallax. Copernicus answered that there was—indeed, there had to be, if the model was correct—such a movement, but the great distance to these stars made the magnitude of the effect very small and therefore not detectable with the instruments then available. This was indeed the reason for assuming that the stars were very far away, but it also showed the scientific method at work, since a testable prediction was unavoidably being made. In fact, Peurbach was the first to attempt to measure the parallax of Comet Halley in 1456. He tentatively concluded that the parallax was less than that of the Moon, and thus that the comet was more distant than the Moon, but his method was highly inaccurate.

Copernicus's theory only appeared in print and in full detail at the very end of his life, though it was widely known at least a decade before its publication, including in Rome. Its publication was to some extent due to the efforts of Georg Joachim Rheticus (1514–1576), a young professor of mathematics and astronomy at the University of Wittenberg. In May 1539 Rheticus arrived at Frauenburg, and early on in what turned out to be a two-year visit he persuaded Copernicus to allow him to publish a brief description of the full model. This came to be known as the *Narratio Prima* (or, to give it its full title, *First report to Johann Schoner on the Books of the Revolutions of the learned gentleman and distinguished mathematician, the Reverend Doctor Nicolaus Copernicus of Torun, Canon of Warmia, by a certain youth devoted to mathematics*). Rheticus was a Protestant, so he took something of a personal risk by visiting a Catholic area and publishing an account of Copernicus' work.

As a small but important point, it is worth noting that the original title of Copernicus' book was *De revolutionibus Libri VI*, or *Six Books on the Revolutions*. Rheticus took the manuscript to the printer Johann Petreius in Nurnberg, but being unable to stay to supervise the printing he assigned the task to Osiander, a Lutheran theologian as well as the town's Lutheran minister. Osiander manifestly took the opportunity and decided to write a letter to the reader *On the hypotheses of this work*, which he used as replacement for Copernicus's original Preface. The letter was unsigned, and for a long time it was thought to be due to Copernicus himself until Kepler revealed the true author 50 years later. Osiander also made a subtle but important change in the title, to *De revolutionibus orbium coelestium*, probably on the grounds that such a title could reasonably be interpreted as meaning the Earth was not necessarily included.

Osiander's text is reproduced in full below, and it manifestly attempts to convey the message that the results of the book were not intended as a true physical model, but a simpler computational method to calculate the positions of the heavenly bodies. This highlights the difference between astronomy and physics (or between mathematics and philosophy, in contemporary language), as they were seen at the time.

To the Reader
      Concerning the Hypotheses of this Work
      There have already been widespread reports about the novel hypotheses of this work, which declares that the earth moves whereas the sun is at rest in the centre of the universe. Hence certain scholars, I have no doubt, are deeply offended and believe that the liberal arts, which were established long ago on a sound basis, should not be thrown into confusion.
      But if these men are willing to examine the matter closely, they will find that the author of this work has done nothing blameworthy. For it is the duty of an astronomer to compose the history of the celestial motions through careful and expert study. Then he must conceive and devise the causes of these motions or hypotheses about them. Since he cannot in any way attain to the true causes, he will adopt whatever suppositions enable the motions to be computed correctly from the principles of geometry for the future as well as for the past.
      The present author has performed both these duties excellently. For these hypotheses need not be true nor even probable. On the contrary, if they provide a calculus consistent with the observations, that alone is enough. Perhaps there is someone who is so ignorant of geometry and optics that he regards the epicycle of Venus as probable, or thinks that it is the reason why Venus sometimes precedes and sometimes follows the sun by forty degrees and even more. Is there anyone who is not aware that from this assumption it necessarily follows that the diameter of the planet at perigee should appear more than four times, and the body of the planet more than sixteen times, as great as at apogee? Yet this variation is refuted by the experience of every age. In this science there are some other no less important absurdities, which need not be set forth at the moment. For this art, it is quite clear, is completely and absolutely ignorant of the causes of the apparent nonuniform motions. And if any causes are devised by

(continued)

the imagination, as indeed very many are, they are not put forward to convince anyone that they are true, but merely to provide a reliable basis for computation.

However, since different hypotheses are sometimes offered for one and the same motion (for example, eccentricity and an epicycle for the Sun's motion), the astronomer will take as his first choice that hypothesis which is the easiest to grasp. The philosopher will perhaps rather seek the semblance of the truth. But neither of them will understand or state anything certain, unless it has been divinely revealed to him.

Therefore alongside the ancient hypotheses, which are no more probable, let us permit these new hypotheses also to become known, especially since they are admirable as well as simple and bring with them a huge treasure of very skillful observations. So far as hypotheses are concerned, let no one expect anything certain from astronomy, which cannot furnish it, lest he accept as the truth ideas conceived for another purpose, and depart from this study a greater fool than when he entered it. Farewell.

Copernicus' reluctance to publishing his full results is somewhat puzzling, but it was probably a combination of his constant efforts to improve the model and a desire to avoid controversy. It has to be remembered that he was not an obscure astronomer working in a lost corner of Europe, but had a solid international reputation. When the Fifth Lateran Council tried to improve the Julian calendar, the Pope asked several experts for advice in 1514, one of these being Copernicus. Instead of travelling to Rome as others did, he simply replied by letter, sending his apologies and further explaining that he felt that the motions of the heavenly bodies were still not understood with sufficient precision for the calendar to be reformed.

Copernicus is said to have received a first complete copy of the printed book on his deathbed (he died of a brain haemorrhage). He clearly didn't see his work as a revolutionary break with the ancient world view—which is how it is often portrayed nowadays—but as its continuation. The text mentions several classical precursors, and in his dedication of the work to Pope Paul III he explicitly mentions the impulse to go back to the original Greek sources to find clues for his astronomical ideas. Astronomers of the sixteenth century regarded the addition of all these complicated circular motions as a refinement of the old theory. It is thus somewhat paradoxical that it is to the title of the book (and to the common but misguided perception thereof) that we owe the modern connotation of words like 'revolution' or 'revolutionary'.

In *De revolutionibus* Copernicus offers several reasons why it is logical that the Sun would be at the centre of the Universe. These were to some extent similar to those that the Greeks could have provided. In particular, he points out that the true motions should be uniform, so if they appear to us irregular that must be due to differing axes of rotation and/or because the Earth is not at the centre of the model's circles. His arguments against Ptolemy were mainly philosophical, and the new theory wasn't the result of improved observations: there were no new empirical facts that might require previous concepts to be abandoned. In fact, Solar System observations could only show motions of other bodies relative to the Earth, and these were the same in both systems; a true observational test would require observations beyond the Solar System, such as the previously mentioned parallax. Its strength was its relative simplicity, which was computationally welcome.

Copernicus' cosmology had a motionless Sun, not at the centre of the Universe, but close to it, and set the Earth in motion. So if one wants to be rigorous, it's not a heliocentric model, but rather a heliostatic one. The main problem with it is that he assumed that all motions were circular and therefore, like Ptolemy, he was forced into using epicycles, though he removed the philosophically more problematic equants. He still spoke of the sphere of fixed stars as a real object, immovable and encompassing everything, but its original function was dissolved into nothing since the stars at rest in space no longer needed to be connected by a material sphere. As a model, his was considered implausible by most astronomers and natural philosophers until the middle of the seventeenth century.

In passing, one should note an important contribution from the English mathematician Thomas Digges (1543–1595). In his work *A Perfit Description of the Caelestial Orbes*, first published in 1576, Digges explained and commented on the Copernican system, but he also introduced a new cosmological development. For a fixed Earth and a rotating sky one automatically assumed a finite, material sphere containing the stars, for how else could they rotate as a whole every 24 h? But once this daily motion was explained by the rotation of the Earth this rotating sphere was no longer needed since the stars could be at rest in space. Moreover, the stars could now be at any distance, and—if you will excuse the pun—the sky no longer had a limit. Copernicus still referred to this sphere as a real object, immovable and containing everything in the Universe. It was Digges who removed the sphere and distributed the stars in an unbounded region. By the end of the sixteenth century several editions of Digges' book had been published, and this pushed the heliocentric Universe and the notion of an infinite Universe to the forefront of the scientific and philosophical debate. One of the main defenders of a Copernican but infinite Universe was Giordano Bruno (1548–1600).

Copernicus was esteemed as the man who improved on Ptolemy, but this was due to the improved numerical values for the model's free parameters, which led to predicted positions of the celestial motions in better agreement with the observed ones. But the model was not seen as a new description of the Universe—it was only a more useful cookbook for calculating planetary positions. Indeed, when the calendar was eventually reformed later in the century and the Gregorian calendar replaced the Julian in 1582, the reform was based on computations using the Copernican model.

Copernicus actually restored the philosophically preferable uniform circular motions, at the cost of putting the Sun in the centre: he removed equants but maintained epicycles. In this sense Copernicus was in fact trying to be a better Aristotelian. He was no revolutionary—he had both feet firmly in the classical tradition. He did not want to change the global structure of the system, he simply changed the location of some of the objects within that system. And in abandoning the use of the equant he actually ended up needing more epicycles than were necessary in Ptolemy's model. His model was complicated in mathematical detail but simpler in philosophical principle. For example, there was no fundamental difference between superior and inferior planets (the position of the Earth among the planets explained the different kinds of observed motions), and retrograde motion was automatically explained, as well as the fact that superior planets were brightest when at opposition. And even though the epicycles remained, their role was not to explain retrograde motion, but only to get the varying planetary speeds in better agreement with observations.

Copernicus' heliocentric theory was vulnerable to theological objections (arising from the authority of the Bible) as well as physical objections (from the authority of Aristotle's doctrine and everyday experience). Accepting the Copernican system as a physical model necessarily required abandoning Aristotelian physics. Recall that according to Aristotle bodies could have only one sort of motion at a time, so how could the Earth have several? Additionally, it implied that the stars were at enormous distances from the Earth, presumably leaving huge empty spaces between them and the planets. Such a vacuum was philosophically absurd. And the Aristotelian notion of a natural place was obliterated if the Earth was no longer at the centre—indeed, and even more objectionably, it turned out that there was nothing at the centre—and the same went for the concepts of natural and violent motion.

Interestingly, the theological objections were initially stronger for Protestants than for the Catholics. Several cardinals had encouraged Copernicus to publish his work, and a pope accepted the dedication of his book, while Protestant leaders like Luther and Melanchthon sharply rejected it. For Protestantism the strict literal validity of the Bible was the basis of faith, whereas the Catholic Church allowed the possibility of interpretation, as long as this interpretation was properly done by theologians.

In the long term, however, the result turned out to be opposite. Protestants, who gradually gained some freedom in interpreting the Bible personally, accepted heliocentrism more quickly. Catholics became more cautious once the Counter-Reformation got started. The culmination of this process was the 1616 declaration by Pope Pius V that the Earth was at rest, and the heliocentric model heretical. At this point, Copernicus' work was put in the index of forbidden books, and it remained there until 1822.

3.4   Tyge Brahe

While Copernicus laid the first stone of modern theoretical astronomy, the Dane Tycho Brahe (1546–1601, see Fig. 3.2) did the same for modern observational astronomy. He was the last great astronomer not to use a telescope. Being of noble origin, he used his wealth to build, in the island of Hven, which he received from Frederick II (1534–1588) in 1576, what was by far the best observatory of his time. As in Peurbach's case this was a full-scale research institute, including numerous visiting scholars, instrument workshops, a printing press, and housing for all the staff and visitors.

Tycho (the Latinised version of his first name Tyge) was born on 14 December 1546 in Knudstrup (Denmark), and died on 24 October 1601 in Prague, Bohemia (now the Czech Republic). He was one of twin sons, but his twin died shortly after birth. His parents had one older daughter but Tycho was their eldest son. He first studied at the University of Copenhagen, where he developed his interest in astronomy, having been particularly impressed by the eclipse of 21 August 1560, and more specifically by the fact that the planetary positions could be known so accurately that the circumstances of eclipses could be predicted well in advance. A scholarly life was unusual for someone of his social rank, nobility just below the royalty. For example, Shakespeare's Hamlet includes two nobles, Rosenkrantz and Guildenstern, who were Tycho's great-great-grandparents.

He continued his studies at the University of Leipzig and the astronomy professor there, Bartholomew Schultz, taught him how to improve the accuracy of observations, and the importance of good instruments. At age seventeen, he wrote:

Fig. 3.2 A 1586 portrait of Tycho Brahe, framed by the family shields of his noble ancestors, by Jacques de Gheyn (Public domain image)

> I have studied all available charts of the planets and stars and none of them match the others. There are just as many measurements and methods as there are astronomers and all of them disagree. What is needed is a long term project with the aim of mapping the heavens conducted from a single location over a period of several years.

Later, he visited the universities of Wittenberg and Rostock. While in Rostock he fought in a duel in which part of his nose was cut off. The duel was not for any of the usual reasons, but for a very different (but perhaps equally important) point of honour: who was the better mathematician. For the rest of his life Tycho had an artificial nose made of silver and gold.

King Frederick of Denmark offered Tycho the island of Hven (called Ven today), together with its tax revenues. Thanks to Tycho, at this time Denmark was spending about 8% of its national income on astronomy—a number that has never subsequently been matched. In 1576 Tycho started building an observatory, which he dubbed Uraniborg. This had exceptionally large and accurate instruments, which he used to collect twenty years' worth of exquisite astronomical observations of the planets and stars, whose quality was only improved many decades later. His star catalogue had better than arcminute accuracy, which is effectively the physical limit for naked-eye observations. It was Tycho's data (and particularly that on the planet Mars) that some years later enabled Kepler to realise that the planetary orbits were ellipses and not circles.

Tycho was aware of the fact that one way to obtain more accurate observations was to build larger instruments, but also knew that this would come at a cost, and not just a financial one. The size of instruments would itself introduce some errors, since for example their weight could deform their various components. In building large instruments it was therefore necessary to make them as stable and robust as possible, to minimise those errors. Increasing the stability and general quality of the instruments also increased the confidence in the data. Additionally, he was careful to estimate and specify the limits of accuracy of each instrument, to calibrate each one of them, and to determine their margins of error.

Tycho played a key role in dismissing the Aristotelian doctrine of the perfect, eternal and unchanging nature of the celestial spheres. Previously, the occasional appearance of comets would be explained in an Aristotelian context by saying that they were simply atmospheric phenomena, just like rainbows or shooting stars, and therefore part of the sub-lunar world. On the evening of 11 November 1572, Tycho noticed an extra star in the constellation of Cassiopeia, as bright as Venus: it was in fact a supernova, now usually known as Tycho's supernova. His observations of it conclusively proved that the star was not a sub-lunar phenomenon, though it should be noted that similar (but rather less quantitative) arguments had also been made regarding previous comets earlier in the century. Several important astronomers of the time accepted the supra-lunar nature of the supernova, including Maestlin (whom we will mention again soon) and Christopher Clavius (1538–1612), a Jesuit astronomer responsible for completing the Gregorian calendar reform.

The proof involved the concept of parallax in a slightly different form, namely diurnal parallax (due to the Earth's rotation), rather than annual parallax (due to the Earth's motion around the Sun). Tycho measured the new star's distance from the pole star and nearby stars in Cassiopeia, both when it stood near the zenith and when, 12 h later, it stood low in the north beneath the pole. If the star had been at the same distance as the Moon, it should have had a parallax of 1 degree and therefore it would have appeared in the second case one degree lower relative to the other stars in Cassiopeia. But in both cases the distances were found to be the same within his observational uncertainty (a few arcminutes). The conclusion was that the new star had no perceptible parallax and was at a far greater distance than the Moon. Contrary to Aristotle's view, changes did occur in the world of the stars. Tycho published a book on the new star in May 1573, having overcome his hesitation as to whether it was proper for a nobleman to do so. He apparently decided to do this after he had seen the mass of nonsense that was being written and published about the star.

Among the most exciting of Tycho's Uraniborg observations was a comet which he first saw on 13 November 1577 (see Fig. 3.3). He published his account in *De mundi aetherei recentioribus phaenomenis* (1588), again showing how his parallax measurements implied that the comet was not closer to the Earth than the Moon, contradicting Aristotle's model of the cosmos. Similar results were obtained by Thaddaeus Hagecius (1525–1600) in Prague and Cornelius Gemma (1535–1578) in Leuven, while Maestlin could not detect any parallax. But Tycho went further, showing that the comet was certainly further away than Venus, and he conservatively concluded that the comet was at least 6 times farther away than the Moon, despite the fact that it was passing effortlessly through the supposed invisible spheres that were thought to carry the planets. The implications were clear: the heavenly spheres did not exist after all. But if planets were not attached to any spheres, how did they move?

Tycho's copy of *De Revolutionibus* survives with annotations—he clearly admired the maths but couldn't accept the heliocentric model. In fact he initially accepted Copernicus' model, but later changed his mind, The problem was, again, related to parallax: he fully understood that a parallax shift should

Fig. 3.3 The Great Comet of 1577, seen over Prague on 12 November. Engraving made by Jiri Daschitzky (Public domain image)

be observed in the Copernican model, but despite more than two years of attempts, particularly focusing on the planet Mars, he was unable to detect it. It was in the course of this effort that he accumulated data that would be vital for Kepler.

Being a firm believer in the statement that a theory must be supported by experimental evidence, he saw only two possibilities to explain this unsuccessful attempt at detection. The first was that the heliocentric model was incorrect and the Earth was in fact fixed. The second was that the effect existed but was too small to detect, even for his exquisite instruments. However, for this to happen the scale of the Universe would have to be unbelievably large. Tycho's measurement accuracy was better than 2 arcminutes, and this meant that his failure to detect the annual parallax implied that the stars must be at least 700 times further away than Saturn. But he deemed the possibility of such a large void useless and nonsensical.

Of course, today we know that the second explanation is actually the correct one, and indeed the stars are so far away that Tycho would have had no hope of measuring this parallax with his instruments. The first measurement of the parallax of a star was only achieved in 1838, the measured angle being about 100 times smaller that Tycho's observational errors. For Tycho the physical objection was the decisive one, but the theological argument was also important: an opinion contrary to the Bible would hamper his work of renovating practical astronomy.

So Tycho rejected the Copernican model, arguing that it failed to answer Ptolemy's objections. Nevertheless, he could not completely ignore the heliocentric arguments, so around 1583 he produced a compromise hypothesis in which the planets moved around the Sun, but the Sun (and also the Moon) moved around the Earth, which was still at the centre of the Universe. This is depicted in Fig. 3.4. This preserved the bulk of Aristotelian physics: the Sun and planets might easily circle the Earth, being made of the fifth element rather

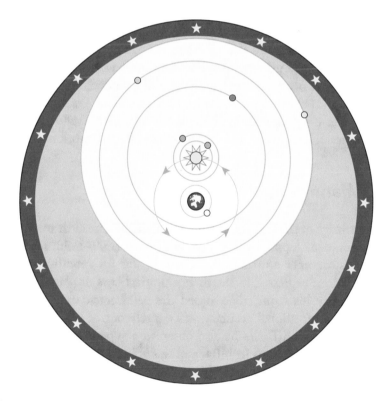

Fig. 3.4 A cartoon view of Tycho's model. The objects on blue orbits (the Moon and the Sun) orbit the Earth, while those on orange orbits (Mercury, Venus, Mars, Jupiter, and Saturn) orbit the Sun. Surrounding everything is a sphere of fixed stars (Public domain image)

than earthly rock. He also used uniform circular motions and epicycles, but just like Copernicus, removed the objectionable equant.

The Copernican and Tychonic models were fully equivalent in their descriptions of the Sun, Moon, and planets. Only observations of the stars could distinguish them, through a detection of the annual stellar parallax. This Tychonic system played an important historical role, since the Roman Catholic Church adopted it for many years as its official astronomical conception of the heavens following Galileo's demonstration (through the observation of the phases of Venus, which we will soon discuss) that the Ptolemaic model was incorrect.

In 1599 Tycho was appointed Imperial Mathematician to the Holy Roman Emperor, Rudolph II, in Prague. where Johannes Kepler joined him as a mathematical assistant; the two first met on 4 February 1600. Tycho's ultimate aim for the mathematical analysis he assigned to Kepler was to validate his own cosmological model, but he died soon afterwards, and Kepler succeed him as Imperial Mathematician and also inherited the data. It is commonly thought that Tycho died from a urinary infection, but there have also been suggestions that he might have been poisoned, and one of the potential suspects is Kepler himself—who was clearly keen to get his hands on all of Tycho's high-quality data. This data eventually allowed Kepler, who unlike Tycho accepted the heliocentric model, to deduce his three laws of planetary motion.

3.5   Johannes Kepler

As the knowledge of the Copernican model gradually spread through Europe, its most prominent supporter was Johannes Kepler (1571–1630, see Fig. 3.5), a German astronomer and mathematician. It was his modification of the Copernican ideas, expressed in the three empirical laws of planetary motion which now carry his name, that turned the heliocentric hypothesis into a theory whose observational accuracy was entirely out of reach of Ptolemy's model. Just like Tycho Brahe he also observed a supernova, in this case in the constellation Ophiuchus in 1604, and was able to show that it was much further away than the Moon or the planets.

Kepler had a deep belief that Nature was the book where God's plan was written, in such a fashion that divine ideas would correspond to geometric objects (a Platonic concept), and that the human mind was ideally suited to discover and understand this structure. His mathematical approach was rooted in the old Pythagorean belief that the working of the heavens could be explained using numbers and geometry.

Fig. 3.5   Portrait of Kepler by an unknown artist, ca. 1610 (Public domain image)

Apart from his laws of planetary motion, he also did important work in optics, discovered two new regular polyhedra (part of what are now known as Kepler–Poinsot polyhedra), gave the first mathematical treatment of the close-packing of equal spheres and the first proof of how logarithms worked, and devised a method for finding the volumes of solids of revolution that (with some hindsight) can be seen as contributing to the development of calculus.

Among Kepler's various contributions to optics he explained the mathematical principles behind the telescope, only a few months after learning of its existence. This was historically important, because understanding how the telescope worked—and specifically how its images were formed—was a crucial step for arguing that the observations made with the telescope are reliable.

Kepler was born on 27 December 1571 in Weil der Stadt, and died on 15 November 1630 in Regensburg (both now in Germany). He studied at the University of Tubingen and was taught astronomy by one of the most important astronomers of the day, Michael Maestlin (1550–1631). This astronomy would have been the classical geocentric version, but Maestlin was familiar with the heliocentric model and one of its early supporters, and Kepler was one of the few pupils that Maestlin considered competent enough to be introduced to it and its mathematical details. It was also Maestlin who told Kepler that the preface in the *De Revolutionibus* was in fact written by Osiander. Kepler seems to have accepted almost instantly that the Copernican system was physically true.

An open and superficially fundamental problem at the time was to explain what determined the number of planets in the sky. In 1595, while teaching in Graz, Kepler suggested that the number and spacing of the six Copernican planets might be explained by a purely geometric construction. This is described in his *Mystery of the Cosmos* (*Mysterium cosmographicum*, Tubingen, 1596). Still following Aristotle at this time, he assumed that they were moving on the surface of perfect spheres, but imagined that these spheres were inscribed and circumscribed by regular polyhedra, of which there were precisely five—as proved by Euclid, although they are still known as the Platonic solids. Explicitly, his idea was the following—see also Fig. 3.6:

Put an octahedron around Mercury's sphere, and surround it by a new sphere: this second sphere is that of Venus

Put an icosahedron around Venus' sphere, and surround it by a new sphere: this third sphere is that of the Earth

Put a dodecahedron around Earth's sphere, and surround it by a new sphere: this fourth sphere is that of Mars

Put a tetrahedron around Mars' sphere, and surround it by a new sphere: this fifth sphere is that of Jupiter

Put a cube around Jupiter's sphere, and surround it by a new sphere: this sixth sphere is that of Saturn

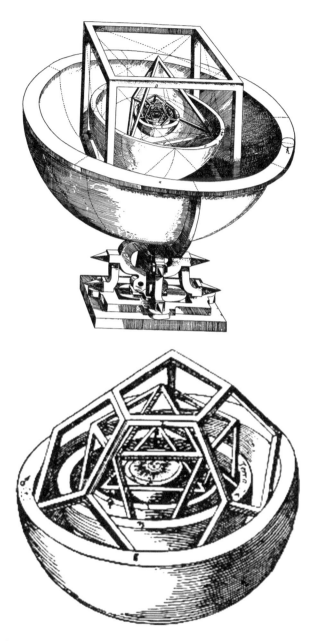

Fig. 3.6 Kepler's Platonic solid model of the Solar System, from *Mysterium Cosmographicum* (1596). The bottom part is a close-up of the inner section of the model (Public domain images)

Kepler saw his cosmological theory as one of the pieces of supporting evidence for the Copernican theory, and he pointed out that another advantage of the heliocentric model over the geocentric one was its greater explanatory power. To give one example, the Copernican theory neatly explained the different apparent behaviour of the inner planets (Mercury and Venus) as compared to the outer ones, whereas in the geocentric theory there was no explanation for this difference.

This hypothesis not only predicted that the planets move in circles but also explained its number (five platonic solids, therefore six planets) and gave numerical predictions for the ratios of their orbits. Initially, he could get distances to agree up to a 5% error, except for Jupiter, something he ascribed to inaccurate measurements. That the arrangement works so well may seem remarkable, but one must keep in mind that one is free to choose all possible combinations of the ordering of the five solids.

In an attempt to obtain better observations and improve the agreement with his model, he sent a copy of the book to Tycho Brahe, but when he later got hold of Tycho's very accurate measurements he was disappointed to find that they disagreed with his hypothesis, and therefore ended up dismissing it. Meanwhile Tycho, then working in Prague, had in fact already written to Maestlin asking for recommendations for a mathematically proficient assistant, and as a result of both events Kepler got the job as Tycho's assistant.

With Tycho's data in hand, Kepler calculated the orbit of Mars using the Copernican system. The circular orbit assumption failed by only 8 arcminutes, one quarter of the diameter of the full moon. Remarkably, he listened to the data and ignored philosophical prejudices. He knew that Tycho's data was accurate to 1 or 2 arcminutes: a model that failed by 8 arcminutes had to be wrong. This was evidence that the orbit of Mars could not be circular, and sometime in 1605 Kepler realised that the orbit was in fact an ellipse.

Thus 3000 years of perfect circular motion were over. Indeed, Kepler wrote that he persisted so long in trying to find a circular orbit for Mars because the necessity of such perfect planetary motion "was taught on the authority of all philosophers, and is consistent in itself with metaphysics". He overcame the philosophical appeal of circles (something that Galileo never did, in fact) only because of his essential commitment to the authority of empirical data over and above the opinion of the philosophers: regardless of other considerations, no theory could survive contradiction by the facts.

An important step in the above analysis (and also crucial to his subsequent work) was his discovery that the orbits of the Earth and other planets are in planes which contain the Sun. Ptolemy and Copernicus both needed somewhat handwaving explanations for the movements of the planets in the direction perpendicular to the ecliptic and for how much they wandered north and south of it, the amount being different for different planets. Kepler realised that these movements were merely a consequence of the fact that their orbits were in planes which were inclined with respect to the plane of the Earth's orbit.

For Kepler, the concept of the epicycle seemed contrary to physics, because the centre of motion was empty, and empty space could not possibly exert any kind of force on a planet. His only objection to the model by Copernicus was the fact that the Sun was close to but not exactly at the centre—in other words, that the model was heliostatic but not quite heliocentric. This highlights the fact that from the beginning Kepler had assumed that the motions of the planets should have physical causes.

Kepler considered the Moon to be similar to the Earth, dark and having high mountains, well before the invention of the telescope. With his interest in optics he explained the faint light of the totally eclipsed Moon as being due to refraction of the Sun's light when passing through the Earth's atmosphere. He also defended the explanation, previously given by Maestlin, that the pale illumination of the lunar disk beside the crescent (the old Moon in the arms of the young, as it is sometimes called) as light reflected upon the Moon by the part of the Earth which is being illuminated by the Sun.

Kepler was keenly aware of the importance of determining the dynamical principle responsible for the planetary motions, which was indeed a pressing need in the new Copernican paradigm. He was not satisfied with the assumption that celestial beings were responsible for it, but was convinced that the motion of all planets was the result of the Sun's dynamical influence. This was in fact his motivation for determining the orbits of the planets to such high accuracy, starting with Mars (whose irregular movements were most noticeable, a result of a combination of its proximity to the Earth and its relatively high eccentricity), and again using Tycho Brahe's data. From this analysis he was able to infer the three simple empirical laws that now bear his name:

Planets move in ellipses with the Sun at one focus

Planets sweep out equal areas in equal times in their motion around the Sun

The average distance to the Sun cubed is proportional to the period squared

Note that although we have been using the term laws, this was not what Kepler himself (or indeed anyone in his generation) actually called them. In fact the term Kepler's laws was only introduced in the eighteenth century. For Kepler they were a set of regularities which described how the Universe seemed to work, and motivated the follow-up question of what kind of force may be behind them.

The first two were published in his *Astronomia Nova* in 1609 (incidentally, the second was actually discovered before the first), while the third was discovered in 1618 and appeared in his *Harmonice Mundi* in 1619. He subsequently showed that these laws applied not only to Mars but also to the other planets, and indeed also to the Galilean satellites orbiting Jupiter. These results therefore dealt a fatal blow to the assumption of circular motion, a cornerstone of the Aristotelian view. The reason why it had persisted for so long was in part philosophical (circles being considered most perfect), but it was also due to the fact that in the Solar System the planetary orbits have a low eccentricity, and therefore circle-based models can do a quite reasonable job in saving the phenomena: the Earth's orbit differs from a circle only by about 1.7%.

Kepler envisaged the Sun as a centre of force in the Solar System. This was reasonable given the fact that the more distant a planet was from the Sun, the longer its period of revolution would be, as expressed in his third law. Before Kepler this was explained by essentially philosophical arguments. Kepler took the bold step of replacing one *anima motrix* per planet with a single *anima motrix* in the Sun, which in fact Kepler referred to as a *vis* (force). Indeed, he can be said to be the person who introduced physics into astronomy.

The specific nature of the Sun's influence remained a mystery to Kepler, though he thought that it should be related to magnetism—a topic of great interest in physics at the time, and the subject of a historically important book, *De Magnete* published in 1600 by William Gilbert (1544–1603), which described the Earth itself as a giant magnet. He did not know (or claim to know) the reason why the planets should follow these laws. In fact, he didn't refer to them as laws at all, but as a set of 'harmonies' that reflected God's design for the Universe—a notion that the philosophers of the Pythagorean school would happily have subscribed to.

Kepler was the first to state clearly that the way to understand the motion of the planets was to understand the nature of the force emanating from the Sun. In contrast to Galileo, Kepler thought that a continuous force was necessary to maintain motion, so he described the force from the Sun as a rotating spoke on a bicycle, continuously pushing the planet around its orbit. Note that neither Kepler nor Galileo fully understood what we now call—for the obvious reason—Newton's first law.

One thing that Kepler got right (unlike Galileo) was that the tides were caused by the Moon's gravity. Kepler's results provided Newton with a guiding light, and the fact that Newton's theory could naturally reproduce Kepler's laws was essential in the acceptance on Newton's theory of gravitation.

The comparative mathematical complexity of Kepler's laws of planetary motion meant that initially only a handful of mathematically competent astronomers recognised Kepler's achievement and its implications. But the tremendous increase in the accuracy of the prediction of planetary positions would in time prove irresistible as an incentive for astronomers to become acquainted with them.

One of the demonstrations of the power of his computational techniques was the prediction that Mercury would transit the Sun on 7 November 1631, while Venus would transit a month later (although unfortunately the second transit would not be visible from Europe). The availability of the telescope, together with projection techniques developed to observe sunspots, ensured that Mercury's transit could now be observed. Kepler published his calculations in 1629, together with instructions for observing the transits. Unfortunately, he did not live to see them confirmed (he died in 1630), but Mercury's transit did occur within 6 h of Kepler's predicted time and several people were able to see it. Astronomers thus had to acknowledge that the observational predictions of Kepler's planetary theory were at least 20 times more accurate than anyone else's.

3.6   Galieo Galilei

The new paradigms introduced by Copernicus, Tycho, and Kepler implied that the key medieval ideas about the heavens were incorrect, and required a re-evaluation of the way the Universe was perceived. Indeed, a moving Earth required a re-evaluation of the theories of motion, which in particular could deal with changing reference frames, so that the behaviour of falling bodies could be reconciled with a rotating Earth. Indeed, why do heavy bodies still fall towards a moving Earth, as opposed to falling towards the Sun, if indeed the Sun is at the centre (or very near it)? This new urge to understand Nature, combined with a period of unprecedented creativity in devising and improving scientific instruments that extended the limited abilities of our senses (most notably, for our purposes, the telescope), led to the birth of modern science.

Contemporary universities were dominated by Ptolemy in astronomy and Aristotle in philosophy. However, Aristotle's theory of motion was such a cumbersome one that it was becoming a major obstacle to scientific progress.

Opposition to Aristotle's theory had been mounting, and although the impetus theory had been smoothing over some of the earlier difficulties it was still quite unsatisfactory.

In the sixteenth century, thanks to the opportunities afforded by the increasing use of artillery in war and of simple mechanical devices in technical work, more attention was paid to mechanics, and opposition to Aristotle increased. In 1585 Giambattista Benedetti (1530–1590) explained the acceleration of falling bodies by the weight continually adding new push to the existing impetus, and he also criticised Aristotle's view that the velocity of falling increased with weight. This was effectively an early statement of what we now call the equivalence principle. In was in this context that Galileo made his contributions to physics and astronomy.

Galileo Galilei (1564–1642, see Fig. 3.7) was born near Pisa on 15 February 1564 and died on 8 January 1642 in Arcetri (near Florence). He was a second-generation Renaissance man, his father being a musician and composer, and part of the circle of musicians and poets that created modern opera. He studied in Pisa and Florence, initially with the aim of becoming a medical doctor, but his real interests were clearly in mathematics. Having completed his studies he held several positions as a mathematics teacher in the universities of Pisa and Padua.

While in Padua, Galileo argued against Aristotle's view of astronomy and natural philosophy in three public lectures (each to an audience of about 300 people) which he gave in connection with the appearance of what is now called Kepler's supernova, in 1604. Galileo used the same parallax arguments that have already been discussed to prove that the new star could not be close to the Earth (as Aristotelians would have argued) but had to be beyond the Moon. He noted that this implied that this region could host stars that came into being and then disappeared, which directly contradicted Aristotelian doctrine. Note that Copernicanism was not the issue here. The debate revolved entirely around the Aristotelian doctrine of the immutability of the heavens. In a letter written to Kepler in 1598, Galileo stated that he was a believer in the theories of Copernicus, but his public statements to this effect only appeared many years later.

Galileo's originality lies not so much in what he found: as we will see shortly, many of the things he found were also independently found by others, and indeed sometimes before he did. It lies principally in how he interpreted his discoveries, understood their relevance, and communicated them to others.

Fig. 3.7   Galileo Galilei, portrait by Domenico Tintoretto, ca. 1605 (Public domain image)

3.6.1 Galilean Physics

While at the University of Pisa, Galileo wrote *De Motu*, a series of essays on the theory of motion which he never published, possibly because he was not fully satisfied with it—indeed, it contains several incorrect ideas. But one new and crucial idea that it also contains is that one can test theories by conducting experiments. He specifically became interested in testing theories about falling bodies using inclined planes. The significant advantage in doing so is that his limiting experimental factor was the uncertainty in measuring the small time intervals associated with typical fall times. Not being able to reduce the absolute uncertainties below a certain value, the alternative was to reduce the relative uncertainties, which in turn involved increasing fall times by slowing down the rate of descent. By using inclined planes with gentle slopes he could effectively dilute the effect of gravity.

It is to Galileo that we owe many of our current notions about motion, such as the law of inertia, the concepts of velocity and acceleration, and the correct description of projectile motion. The crucial mental block limiting the development of planetary mechanics was the belief that circular motion was the natural one. Although Galileo didn't entirely overcome it, he did understand and accept that a straight rather than circular path was the natural one for the motion of a free body.

Galileo also pointed out that all bodies should fall with the same acceleration in a given gravitational field. This concept—sometimes referred to as the universality of free fall—would become a key concept in Einstein's relativity. He pointed out that the Aristotelian theory could not be correct, using a simple theoretical argument. Assume that how fast objects fall does depend on the object's mass. If you take a heavy and a light object and drop them, side by side and from the same height, Aristotle would naturally predict that the heavier one would fall faster. But now drop the two together, maybe attached by a thin rope or wrapped in some light film. Should the heavier one force the lighter one to go faster, or the lighter force the heavier one to go slower? On the other hand, the composite object is now even heavier than the original heavy one, so it should fall even faster. Thus the only way out of this contradiction is if everything falls at the same rate.

Before Galileo almost nobody bothered to check whether bodies of different weights fell with different velocities. As has been mentioned several times already, Aristotelian mechanics held that heavier bodies fell more quickly than lighter ones, although we have also pointed out notable exceptions like John Philoponus (who stated that the differences in the fall times were very small) and Giambattista Benedetti (who proposed equivalence). Also of some note is Simon Stevin, who did test this experimentally in 1586. So although Galileo was definitely not a pioneer in this regard, he did test the idea more systematically, as well as proposing more physical arguments for it. Having said that, he certainly did not conduct the experiment of dropping two bodies from the leaning tower of Pisa—this might have been done by his last student and first biographer, Vincenzo Viviani, who was the one who referred to this in his posthumous biography.

Galileo was unable to find a proof of the Earth's rotation, although he did show that the traditional objections were not valid—most of these came once again from misunderstandings about relative motion. If the Earth moves, its inhabitants share its uniform motion and it therefore remains imperceptible to them. Galileo also clearly recognised that the vertical and horizontal motions of a given trajectory were independent components that could be studied separately. This may seem trivial to us but was in fact a major conceptual advance, given the Aristotelian emphasis on the fact that each object could only have one type of motion.

In Aristotelian physics, circular motion in the heavens was natural and needed no explanation, and the same was true for vertical motion towards the Earth. Galileo showed that it was instead uniform motion or rest that was the natural state, and that changes in the state of motion were what needed to be explained. This is often referred to as Galilean inertia. A force causes a change of state (acceleration), not motion per se. Galileo, however, never fully understood the distinction between accelerated and unaccelerated motion, and still thought of natural motion as a natural instinct, further regarding uniform circular motion as inertial. An objection to the diurnal motion of the Earth, asking what causes this motion, was solved by stipulating that the Earth had a natural tendency to revolve around the centre of its mass once every 24 h.

Galileo came close to but never quite formulated Newton's first law of motion, because he had to make circular motion a cornerstone of his heliocentric system in order to answer the objections of his opponents. He described circular motion as a balance between force and resistance, and believed that circular motion was a natural state that would persist unless acted upon by some external agent. This brought Galileo close to modern physics, but he never formulated the correct principle of inertia because he was still thinking in terms of an eternally ordered motion. In other words, his terrestrial and celestial mechanics were not fully inertial.

He did still think in terms of the impetus acquired as a body falls, and he continued to speak of intrinsic motions, both of which were later banished from inertial mechanics. Galileo did not regard gravity as an external force field, but as an intrinsic property of a body. His circular motions were always made up of two components, one of them being a rectilinear tangential tendency to persist in the line of the impetus. Because uniform circular motion was seen as natural, Galileo did not need a force acting on the planets to keep them orbiting, provided their orbits were circular (while for an off-centre elliptical motion, one would most certainly need a central force).

This shows that separation of a particular motion into inertial and noninertial components, which we learn in high school and university physics and take for granted today, is more subtle than it may appear. Galileo's concept of inertia was not quite correct, and our modern idea of inertia is due instead to

René Descartes and Christiaan Huygens (1629–1695). Galileo's main physics legacy was the demonstration that the barrier that precluded the presence of two natural motions in one body was no more than a mental block. But Galileo's Universe still had no gravity, unlike Kepler's.

3.6.2 Galilean Astronomy

In 1609, probably in May or June, Galileo heard from his friend Paolo Sarpi (1552–1623), a friar interested in science, of an instrument made by Hans Lippershey and shown in Venice that allowed the observer to see distant objects with relative clarity. He set about improving it, and when he finally received a copy of it he found out that the ones he had meanwhile made were already better.

It's not hard to make a small telescope using spectacle lenses: take a weak plano-convex lens as the objective and strong plano-concave one as the eyepiece, put them together in a tube of suitable length, and you can easily get a spyglass that will magnify three or four times. Who the original inventor of the telescope was remains a mystery. The first written description of the use of a telescope appears in the French pamphlet *Le Mercure Francois* in December 1608. Girolamo Sirtori, in his *Telescopio Sive Ars Perficiendi* (1618) suggested that the inventor was Hans Lippershey (1570–1619), a spectacle maker in Middelburg, but other candidates include Jacob Metius from Alkmaar and Sacharias Janssen (1585–1632), who actually lived next door to Lippershey. The three applied for patents in the space of less than three weeks (Lippershey on 25 September, Jansen on 14 October, and Metius on 15 October 1608), but this was not granted to any of them.

Galileo was not the first to use a telescope for astronomy. For example, in England Thomas Harriot (1560–1621) was mapping the Moon as early as 5 August 1609. His first Moon map was made with a telescope which had a magnification of six times, as he records on the drawing. Harriot also observed sunspots and Jupiter's moons. But Galileo also quickly realised its astronomical potential and turned it toward the heavens. By August 1609 he had reached a magnification of 8 times, which was better than anything made by others at the time. By January 1610 he was up to 20 times, and later that year he eventually reached a magnification of more than 30 times. That said, note that for many of his observations it was not so much the magnification as the light gathering and resolving power of the telescope that allowed him to see what nobody had seen before. Galileo began his career as an astronomer on the night of 30 November 1609. The name 'telescope' was coined on 14 April 1611 by a Greek-born mathematician living in Rome, Giovanni Demisiani.

The telescope resolved the Milky Way into many thousands of stars, and revealed that the familiar constellations contained very many members that were invisible to the naked eye. The discovery of so many new stars prompted Kepler, in his *Dissertatio cum Nuncius Sidereo* (1610) to discuss why they should be so faint, and thence why the sky was dark at night—an early discussion of what we now know as Olbers' paradox, Kepler's *Dissertatio* can therefore be considered to mark the birth of observational cosmology. But it was Galileo's observations of the Solar System that would have a more dramatic impact.

Observing the Moon's surface, Galileo noticed that the line between the dark and illuminated sides, called the terminator, was twisted and irregular, rather than a straight line. He also noticed that the surface was heavily scarred, and identified some dark features as shadows. This meant that, like the Earth, the Moon was full of mountains and valleys: it was not exactly spherical, and certainly not perfect. Galileo even attempted to measure the height of some of the lunar peaks, and his estimates were rather accurate given the means available to him—he actually underestimated heights by about 30%.

Observing the secondary light of the Moon (the faint illumination of the dark side of the Moon in its crescent phase), he guessed that this was due to sunlight reflected by the Earth, as had previously been suggested by Michael Maestlin. More than 25 years after his first observations, and just before losing his sight, Galileo also noticed that it was occasionally possible to see small areas that do not usually form part of the visible area, and he correctly attributed this to a small 'rocking' motion of the Moon when seen from the Earth, which he called libration—a term that is still used. This is due to the fact that the Moon's orbit is an ellipse, and its axis of rotation is not perpendicular to the plane of its orbit. The end result is that over a sufficiently long time one can see a total of 57% of the Moon's surface.

These results showed that the Earth and the Moon were similar: solid rocky bodies with rough surfaces, shining by reflecting light from the Sun and presumably made of the same materials. And if the Earth-like Moon could move around the Earth, then why couldn't the Earth move around the Sun? While other discoveries by Galileo were more important in an astronomical sense, none caused more hostility and attempts at refutation than his Moon results. When in the spring of 1611 Galileo visited Rome he discussed his observations with the Jesuit astronomers, including Clavius who, although quite elderly (he would die early in 1612), was probably the most respected astronomer in Europe. The Jesuits, who by then had their own telescopes and were making observations themselves, confirmed all of Galileo's discoveries, with the proviso that the Moon's surface was not rough but made of denser and rarer parts.

Who was the first to observe sunspots scientifically remains an open (and indeed controversial) question. Candidates include Harriot and Galileo, though the latter only began their systematic study later. What is clear is that at least three people published their results before Galileo: Christoph Scheiner (1573–1650) then professor in Ingolstadt, and David and Johannes Fabricius (who were father and son). Scheiner suggested that they were hitherto unknown tiny stars close to the Sun. Galileo's analysis was far more detailed, and showed through rigorous geometric arguments that the spots must be on the surface or very near the Sun. He noted that their changing speed and separation as they moved across the solar disk were characteristic of motion on the surface of a sphere, which he interpreted as a consequence of the Sun's rotation (whose period he thereby estimated to be about a month), but erroneously believed them to be clouds.

Although Galileo plainly won the scientific argument, his dismissive writing made Scheiner look foolish. This may have been a miscalculation, because he made an implacable life enemy of Scheiner, who in addition to being a competent scientist was also a Jesuit priest. This was one of a number of events which had the effect of turning the whole Jesuit order, which had initially been quite sympathetic to Galileo, against him, and it would be a key factor in Galileo's subsequent troubles with the Church.

Soon after the publication of Copernicus' *De Revolutionibus*, critics of the heliocentric hypothesis noted that, if all the planets moved around the Sun, then Mercury and Venus, like the Moon, should exhibit phases due to their changing positions relative to the Earth, as well as varying in size as they did so. If Venus was always nearer to the Earth than the Sun was, as in Ptolemy's theory, when observed from the Earth it would always appear as a crescent. If it was always further away than the Sun (which was seldom mentioned but was at least still a possibility conceptually), it would always appear as a disk. Note that, in the Ptolemaic system, the spheres could not intersect each other, so either Venus was always closer to us than the Sun or it was always further away. But if Venus moved around the Sun, as in the Copernican or Tychonic theories, it would have a complete cycle of phases, changing from a small round disk near superior conjunction (when it was behind the Sun, and therefore further away) to a large crescent near inferior conjunction (when it was in front of the Sun).

Naked-eye observations failed to provide confirmation of this effect, and thus provided an apparent argument against heliocentrism. Copernicus himself was aware of this point, but he legitimately claimed that the phases existed but couldn't be seen with the naked eye because Mercury and Venus were much further away than the Moon. In December 1610 Galileo's observations provided a striking confirmation. The phases of Venus proved that it at least revolved round the Sun (and the same might presumably happen also for Mercury), whence the Ptolemaic system was untenable. On the other hand, note that these observations were equally compatible with the Copernican and the Tychonic system. Later, in 1639, Giovanni Battista Zupi (ca. 1590–1650), another Jesuit priest, discovered Mercury's phases.

But the most dramatic of Galileo's observations was the fact that Jupiter had four satellites orbiting it. Again he wasn't the first to observe them: Simon Mayr (1573–1625), better known by the Latinised version of his surname, Marius, is thought to have observed them earlier, and also given them the names we now use: Io, Europa, Ganymede, and Callisto. (Incidentally, Marius also independently observed the phases of Venus around the same time as Galileo.) Nevertheless, the four are today collectively called the Galilean satellites. Another criticism of the Copernican system was that an Earth moving around the Sun every year would be unable to keep the Moon with it. Starting on 7 January 1610, Galileo observed four new stars that were always in the vicinity of Jupiter and moved around the planet, sometimes approaching it and disappearing for a short time, but soon reappearing on the other side. He correctly explained this behaviour by asserting that they were satellites orbiting Jupiter, just as the Moon orbited the Earth, and that their vanishings occurred when they moved behind it (as seen from Earth). Jupiter had therefore no trouble keeping its moons in tow, and everyone agreed that Jupiter itself was orbiting something, be it the Earth or the Sun.

Galileo thought that the main objection of the anti-Copernicans was that it would be impossible for the Moon to have a double movement, on the one hand around the Earth and on the other hand with the Earth around the Sun. But the satellites of Jupiter certainly did have a double movement, so that objection was invalid. That meant that there was certainly more than one centre of motion in the Solar System, which flatly contradicted Aristotle. And that being the case, why could a moving Earth not keep its Moon in tow? While this certainly did not prove the Copernican system, it did make it more plausible.

Later, Galileo determined the orbital periods of the four satellites and realised that the innermost satellite was the quickest to go round Jupiter, and the outer satellite was the slowest. This was therefore nothing less than a miniature model of the Solar System as described by Copernicus. Unquestionably, these satellites did not orbit the Earth, and instead Jupiter happily kept not one but all four satellites orbiting it. The determination of these orbital periods was in fact Galileo's most important contribution to mathematical astronomy.

Fear of being scooped induced Galileo to publish his results quickly. By 13 March the *Sidereus Nuncius* (*Sidereal Messenger*) appeared in Venice, reporting on his observations up to 2 March. A total of 550 copies of the first edition were printed, and quickly sold out. A few weeks later Galileo's discoveries were known through most of Europe, and within a year he was the most celebrated natural philosopher in Europe.

All this was in violent contradiction with the Aristotelian model, and therefore gave credence to heliocentrism. From this point onward Galileo publicly defended heliocentrism, and his prestige together with the fact that he wrote and published in Italian (more accessible to the public at large than the scholarly but considerably rarer Latin) made him an obvious target for the ire of the Catholic Church when, in 1616, Pope Pius V declared the Earth to be at rest and heliocentrism heretical. Nevertheless, geocentrism was doomed: within 50 years all but the most medieval scholars rejected it. From a theological point of view Galileo could safely argue for the corruptibility of the heavens (because the Bible said as much), but he moved in dangerous waters when he tried to interpret the Bible to fit his Copernican views, which he also soon did.

Since Jupiter had four satellites, it was natural that Galileo should examine the other planets to see if they too had any. The best he could do was to note that Saturn had a rather elongated shape, though he failed to identify the presence of the rings. This would only be done by Christiaan Huygens (1629–1695) in 1655, who also discovered Saturn's largest moon, Titan.

3.6.3 Physics Versus Philosophy

One of the challenges which Galileo had to face was the issue of whether what the telescope saw was real. The Aristotelian view was clear: the unaided sight of the human eye was the only true source of vision of the material world. At this time spectacles were well known and tolerated as a kind of necessary evil for the elderly or those with dimmed vision. Glass lenses were pioneered around 1350 (an example can be seen in the 1352 portrait of Hugh of Provence by Tommaso da Modena) and there are some unconfirmed suggestions that they may have been in use as early as 1287. Concave lenses appeared later, but it was obvious that lenses produced distorted images.

Now one had an instrument with multiple lenses that seemingly showed completely unknown worlds. Did Galileo's findings exist only inside his telescope? Since lenses could alter images, could they also not create them? It should be noted that in order to get high magnification, Galileo's telescopes had extremely small fields of view—any slight movement would put the intended object out of view. Therefore, their use necessitated very precise pointing, and this was something that requires considerable skill, as you will find if you ever have the opportunity to handle a replica of one of Galileo's telescopes. In some sense, Galileo's skill worked against him. It would often happen that he would try to show his findings to other people—including Aristotelian philosophers, if only they would agree to look through a telescope at all—by explaining where they should point the instrument, and they would claim to see nothing, simply because they were not sufficiently skilled.

Although Galileo made multiple major contributions to astronomy, he was not right in every case. When three comets appeared in quick succession in 1618—and they were the first comets to appear after the emergence of the telescope—Galileo uncharacteristically argued that they were close to the Earth and caused by optical refraction, rather than being celestial objects. This had the unfortunate consequence of antagonising the Jesuits, who were in this case on the right side of the argument, and even tried to calculate orbits for the comets. Thus the Jesuits began to see Galileo as a dangerous opponent.

In a letter to his former student and then Benedictine monk Benedetto Castelli (1578–1643), Galileo first argued that the Bible had to be interpreted in the light of what science had shown to be true. The Catholic Church's most important figure at this time in dealing with interpretations of the Holy Scripture was Cardinal Roberto Bellarmino (1542–1621). He saw Jupiter's moons through a telescope in 1611 and at this point seemed unconcerned regarding the Copernican theory. A letter from Bellarmino to the Neapolitan Carmelite priest Paolo Antonio Foscarini (1580–1616), dated April 1515 and reproduced below, agreed that assuming the Copernican model as a computational hypothesis was fine, and that if there were a true demonstration that the Earth moved around the Sun, then one needed to be careful in interpreting that, as the Bible seemed in contradiction to it. But it is well worth going through the argument in full detail:

My Reverend Father,
   I have read with interest the letter in Italian and the essay in Latin which your Paternity sent to me; I thank you for one and for the other and confess that they are all full of intelligence and erudition. You ask for my opinion, and so I shall

(continued)

give it to you, but very briefly, since now you have little time for reading and I for writing.

First I say that it seems to me that your Paternity and Mr. Galileo are proceeding prudently by limiting yourselves to speaking suppositionally and not absolutely, as I have always believed that Copernicus spoke. For there is no danger in saying that, by assuming the Earth moves and the Sun stands still, one saves all of the appearances better than by postulating eccentrics and epicycles; and that is sufficient for the mathematician. However, it is different to want to affirm that in reality the Sun is at the centre of the world and only turns on itself, without moving from east to west, and the earth is in the third heaven and revolves with great speed around the Sun; this is a very dangerous thing, likely not only to irritate all scholastic philosophers and theologians, but also to harm the Holy Faith by rendering Holy Scripture false. For Your Paternity has well shown many ways of interpreting Holy Scripture, but has not applied them to particular cases; without a doubt you would have encountered very great difficulties if you had wanted to interpret all those passages you yourself cited.

Second, I say that, as you know, the Council [of Trent] prohibits interpreting Scripture against the common consensus of the Holy Fathers; and if Your Paternity wants to read not only the Holy Fathers, but also the modern commentaries on Genesis, the Psalms, Ecclesiastes, and Joshua, you will find all agreeing in the literal interpretation that the Sun is in heaven and turns around the Earth with great speed, and that the Earth is very far from heaven and sits motionless at the centre of the world. Consider now, with your sense of prudence, whether the Church can tolerate giving Scripture a meaning contrary to the Holy Fathers and to all the Greek and Latin commentators. Nor can one answer that this is not a matter of faith, since it is not a matter of faith 'as regards the topic' it is a matter of faith 'as regards the speaker'; and so it would be heretical to say that Abraham did not have two children and Jacob twelve, as well as to say that Christ was not born of a virgin, because both are said by the Holy Spirit through the mouth of the prophets and the apostles.

Third, I say that if there were a true demonstration that the Sun is at the centre of the world and the Earth in the third heaven, and that the Sun does not circle the Earth but the Earth circles the Sun, then one would have to proceed with great care in explaining the Scriptures that appear contrary; and say rather that we do not understand them than that what is demonstrated is false. But I will not believe that there is such a demonstration, until it is shown me. Nor is it the same to demonstrate that by supposing the Sun to be at the centre and the Earth in heaven one can save the appearances, and to demonstrate that in truth the Sun is at the centre and the Earth in the heaven; for I believe the first demonstration may be available, but I have very great doubts about the second, and in case of doubt one must not abandon the Holy Scripture as interpreted by the Holy Fathers. I add that the one who wrote, "The Sun also riseth, and the Sun goeth down, and hasteth to his place where he arose," was Solomon, who not only spoke inspired by God, but was a man above all others wise and learned in the human sciences and in the knowledge of created things; he received all this wisdom from God; therefore it is not likely that he was affirming something that was contrary to truth already demonstrated or capable of being demonstrated.

(continued)

Now, suppose you say that Solomon speaks in accordance with appearances, since it seems to us that the Sun moves (while the Earth does so), just as to someone who moves away from the seashore on a ship it looks like the shore is moving, I shall answer that when someone moves away from the shore, although it appears to him that the shore is moving away from him, nevertheless he knows that it is an error and corrects it, seeing clearly that the ship moves and not the shore; but in regard to the Sun and the Earth, no wise man has any need to correct the error, since he clearly experiences that the Earth stands still and that the eye is not in error when it judges that the Sun moves, as it also is not in error when it judges that the Moon and the stars move. And this is enough for now.

With this I greet dearly Your Paternity, and I pray to God to grant you all your wishes.

At home, 12 April 1615.
To Your Very Reverend Paternity.
As a Brother,
Cardinal Bellarmino

The issue was that Galileo thought he had a proof of the motion of the Earth (based on his theory for explaining the tides), but the Jesuits believed—correctly, as it turned out—that his proof was not valid. As was mentioned in the previous section, a correct explanation for the tides had already been given by Kepler.

As a reply, in 1616 Galileo wrote the *Letter to the Grand Duchess Christina*, arguing for freedom of inquiry and vigorously attacking the followers of Aristotle. He further argued for a non-literal interpretation of Holy Scripture when the literal interpretation would contradict facts about the physical world that had been demonstrated by mathematical science. In this letter Galileo explicitly stated that for him the Copernican theory was not just a mathematical calculating tool, but was a physical reality.

The introduction of the telescope raised a difficult problem for the Church. Galileo's use of evidence gained with the telescope implied access to truth, independent of revelation, allowing the individual observer to interpret the book of Nature—which previously had been the exclusive right of the Church. One of the lasting contributions of Galileo was to bring science out of the seclusion of a scientist's room into the forefront of public debate: not only did he defend his ideas and discoveries among his peers, but he publicised and popularised them when appropriate.

The issue was not over the observations themselves (which the Jesuit astronomers were themselves competently doing), but over their interpretations—not the practice but the theory behind them. Thus in 1616 Pope Paul V declared that the propositions that the Sun was the centre of the Universe and that the Earth had an annual motion were foolish, philosophically absurd, formally heretical, and theologically an error of faith.

On 26 February 1616 Galileo was warned not to hold or defend the Copernican theory (it might only be dealt with as a mathematical hypothesis) and admonished to abandon it; he was also forbidden to discuss the theory orally or in writing. Yet he was also reassured by Pope Paul V and by Cardinal Bellarmino that he would not be put on trial or condemned by the Inquisition. On this verdict, on 5 March 1616 the Congregation of the Index suspended and forbade Copernicus' books and similar teachings until they were amended; *De Revolutionnibus* remained on the Index until 1835.

To some extent the arguments were not much more advanced (theologically) than those of Buridan and Oresme, but they were now in a different context— that of the Reformation and counter-Reformation. While superficially the issue was the Copernican system that Galileo defended, the underlying issue was that of intellectual freedom versus authority. Galileo's ambitions also clashed with those of another powerful group, the Jesuits, who had an interest in discrediting a major competitor. Galileo had originally been on good terms with the Jesuits, who confirmed his observations (though sometimes disagreeing about his interpretations of them, as in the aforementioned case of the Moon), but he gradually alienated them due to episodes such as the sunspot controversy and the disagreement over the 1618 comets. Jesuit mathematicians considered that the suspension of Copernicus' book strengthened their position, and the comets episode provided a means for them to assert their superiority.

Galileo decided to publish his views, believing that he could do so without serious consequences from the Church. He began to write his famous *Dialogue Concerning the Two Chief Systems of the World*—Ptolemaic and Copernican, in 1624, but due to ill health only completed it in 1632. The climax of the book is precisely Galileo's argument that the Earth moves, based on his theory of the tides. Jesuit astronomers (including Scheiner, now living in Rome) reacted violently to the *Dialogue*, seeing it as a glorification of Copernicus. The Church quickly confirmed that the book openly supported Copernicanism, forbade the sale of further copies (though there were none left, since the book had quickly sold out) and ordered Galileo to appear in Rome, though again due to ill health he only arrived in Rome in 1633.

What was at stake in Galileo's trial was not Copernicanism: from the point of view of the Church, that issue had been settled in 1616, and it was taken for granted that the theory was false. Instead, his accusation was based on the fact that in defending Copernicus in his book, Galileo had disobeyed the conditions laid down by the Inquisition in 1616. Having been found guilty,

Galileo was condemned to lifelong imprisonment, but the sentence was carried out somewhat sympathetically and it amounted to house arrest rather than a prison sentence. Galileo first lived with the Archbishop of Siena and was later allowed to return to his home in Arcetri, still under house arrest.

Nevertheless, Galileo managed to write a further book, the *Discorsi e Demonstrazioni Matematiche Intorno a Due Nuove Scienze*, which appeared in 1638. The book summarised much of his life's work in physics and was an excellent popular work which completely put an end to Aristotle's doctrines of motion and of physics. Ultimately, the Galilean revolution, in physics as well as in astronomy, stemmed from the realisation that Nature can be directly questioned, and through experiments and observations be made to provide answers that are formerly hidden from view.

4

Classical Physics

In this chapter we will look briefly at pre-Newtonian gravity, before discussing Isaac Newton himself and his seminal role in the apogee of the Scientific Revolution. We will then look into the discovery of the planet Neptune, arguably the greatest success of Newtonian physics and certainly the most fascinating episode in the history of astronomy, and compare it to the non-discovery of Vulcan. Finally, as an example of the role of Newtonian physics in the modern world, we will discuss some of the physics of orbits and of objects in the Solar System, such as artificial satellites, asteroids, and comets.

> What makes planets go around the Sun? At the time of Kepler some people answered this problem by saying that there were angels behind them beating their wings and pushing the planets around an orbit. As you will see, the answer is not very far from the truth. The only difference is that the angels sit in a different direction and their wings push inwards.
>
> Richard Feynman (1918–1988)

4.1 Pre-Newtonian Gravity

After Galileo, astronomy evolved at a much faster pace. One of the causes of this change was the separation between science and religion, which was accepted by most intellectuals following the Galileo affair. One example of this new generation of intellectuals was René Descartes (1596–1650), who in some sense bridged the gap between Galileo and Newton. Descartes' modern

© Springer Nature Switzerland AG 2020
C. Martins, *The Universe Today*, Astronomers' Universe,
https://doi.org/10.1007/978-3-030-49632-6_4

reputation is based on his philosophy and mathematics, but his work in the field of physics was also highly significant. He introduced mathematical concepts such as space and movement, which proved crucial for subsequent developments in astronomy and cosmology.

Descartes also compared the concepts of matter and space, arguing that were it not for the fact that matter can move, it and space would be identical entities and the Universe would be completely uniform. This led Descartes to the conclusion that the differences between various regions of the Universe had to be the result of movements of matter. In practical terms, one of the emerging requirements was that any movement of matter had to be compatible with the movements of the surrounding matter. Thus movements which otherwise would have been rectilinear would, when taken together, constitute what he called vortices. The Solar System would be an example of such a vortex, consisting of tiny particles whose movements would make them invisible, and it was this material that would carry the planets along in their orbits. Moreover, a centrifugal effect would make some of this matter—which the human eye would see as luminous—move to the centre of the vortex, where it would gradually accumulate and in due course form the Sun.

The Sun was therefore a normal star surrounded by further vortex matter in which the planets moved on their orbits. Descartes also assumed that the vortex around a star could also collapse and become a further star which could move to a neighbouring vortex, but it could also become a planet in the new vortex, or possibly a comet. Even though this model could explain, in a qualitatively acceptable way, the formation of the Solar System with the celestial bodies known at the time, it could not in any way predict how each of them would evolve in the future. The ability to accurately calculate future positions of celestial bodies only emerged with Newtonian mechanics.

By the middle of the seventeenth century the importance of Kepler's laws as a computational tool was widely recognised. They explained the general properties of the apparent motions of the planets on the sky, but on the other hand they could not explain why the planets had these trajectories. What was lacking was a physical mechanism for the forces that were behind these movements. Kepler himself had already searched for this mechanism, but had been unable to find it.

The first explicit statement of the modern principle of inertial relativity was due to Pierre Gassendi (1592–1655), ca. 1630. Gassendi is most often remembered today for reviving the Greek doctrine of atomism. In the 1630s he repeated many of Galileo's experiments with motion, described gravity as an external force, and noted that what he called the natural states of motion were characterised not only by uniform speeds (as Galileo had already said), but also by rectilinear paths. Descartes also described this in his *Principia Philosophiae* (1644), formulating what is now known as Newton's first law of motion, as did Christiaan Huygens (1629–1695) in his *The Motion of Colliding Bodies* (written in the mid-1650s but not published until 1703).

Huygens was the first to apply the Galilean ideas of motion and inertia to the planets. Assuming that the undisturbed planetary orbits would be straight lines, he calculated the radial force that would be necessary to keep a planet in a circular orbit. It was known that when a stone was moving at the end of a rope, its circular path seemed to result from the balance between the force exerted by the person holding the rope and the external force with which the stone pulled the rope. Not later than 1673 Huygens postulated that, for circular planetary orbits obeying the rest of Kepler's laws, the central force had to vary as the inverse of the square of the distance. He first showed that the external force had to be proportional to the square of the velocity and inversely proportional to the radius, but assuming that the orbit was circular one could use Kepler's third law to eliminate the velocity dependence and end up with the inverse square law for the external force. One could also show that Kepler's second law required a central force, and the first law could only be true if the force varied as the inverse of the distance squared.

This result was also known to three British scientists, Christopher Wren (1632–1723), Robert Hooke (1635–1703), and Edmund Halley (1656–1742). They all knew each other and are even known to have discussed the subject, so their results are unlikely to be independent. Between 1662 and 1664 Hooke tried to show experimentally that the force exerted by the Earth on any body varied with height, by measuring the weight of various bodies at several heights above the ground, but he failed to detect any variations.

But the crucial question was a different one: what would be the trajectory of a planet subject to a force varying as the inverse of the square of the distance? Would that necessarily (or indeed at all) be an elliptical orbit compatible with Kepler's laws? This was a more difficult question, because the mathematical tools necessary to answer it did not exist until Newton tackled the problem.

4.2 Newton's Laws

Isaac Newton (1643–1727, see Fig. 4.1) was the most influential scientist of his time, and completely dominated the development of the exact sciences for the next two and a half centuries: his preeminence was only matched by Einstein. Newton provided a simple set of rules—which we now call laws—that describe

Fig. 4.1 Portrait of Newton, aged 46, by Godfrey Kneller, 1689 (Public domain image)

the way in which objects interact gravitationally, and the magnitude of this achievement lies precisely in its accuracy, simplicity, and universality. This produced a qualitative change in the way not only scientists but even the general public viewed Nature: we inhabit a Universe that is ruled by laws, where a known action will produce predictable effects, and not a world filled with careless and whimsical spirits who can intervene in the world as they please.

Newtonian mechanics is an adequate and sufficient tool to describe almost all situations we experience in our everyday lives, from billiard balls, bridges, and buildings to airplanes and space ships like Apollo XI. The only gadget you may have encountered for which Einstein gravity is definitely needed is the GPS, for reasons that we will explain later. So successful was Newton's description of Nature that by the end of the nineteenth century many physicists believed that, apart from a few missing details, our understanding of Nature was basically complete. At the beginning of the twentieth century, however, it was discovered that Newtonian gravity also has its limits of applicability: it fails when describing behaviour at the atomic level or below (which eventually led to quantum mechanics), motion close to the speed of light (which led to special relativity), or the physics of very massive and compact objects (which led to general relativity).

Newton was born on 4 January 1643 in Woolsthorpe, and died on 31 March 1727 in London. Note that the calendar in use in Britain at the time of his birth was still the Julian calendar, and according to that he was born on Christmas Day 1642—the date of 4 January 1643 is the equivalent Gregorian calendar date. Newton came from a family of farmers and studied at Trinity College, Cambridge. His scientific genius revealed itself when the plague closed the University in the summer of 1665, forcing him to return home. There, in a period of about eighteen months—and while he was still under 25 years old—he began revolutionary advances in mathematics, optics, physics, and astronomy.

Almost two decades later, in August 1684 Halley visited Newton in Cambridge. Their conversation touched on the question of what kind of curve would be described by the planets supposing that the force of attraction towards the Sun were proportional to the inverse of the square of their distance from it. Without hesitation, Newton replied that he knew that it would be an ellipse, since he had calculated it. Newton was unable to find his calculations but promised to send them to Halley later. This would become the germ of Newton's masterpiece, and in fact Halley made two further important contributions to it, first by acting as editor and then by financing the publication himself.

Newton's argument began more or less along the same lines as Huygens', by considering the rate at which the Moon must fall towards the Earth in order to remain in its circular orbit and asking what centripetal acceleration would be required to produce this motion. He found, by comparing with the known value of the acceleration of gravity at the Earth's surface, that the gravitational acceleration produced by the Earth at the distance of the Moon decreased as the inverse square of the distance from the centre of the Earth. He used this result not (as is sometimes claimed) to show that the force of gravity varies as the inverse of the square of the distance—which he already knew—but rather to conclude that the required inverse squared force was due to Earth's gravity alone and no extra forces were needed. In other words, Earth's gravity was the only physical mechanism necessary to provide the proper rate of fall of the Moon. He then generalised this to the law of universal gravitation.

Newton was aware that one difficulty in the analysis came from the fact that the Earth and the Moon are extended bodies, and not idealised point particles. Using the new mathematical tool of calculus, which in the meantime he had also developed, he proved that this was not a problem. An interesting property of Newtonian gravity is that, outside a uniform spherical mass, the gravitational force it produces is the same as that of a point particle of equal mass located at the centre of the sphere.

Newton's three laws provide the means to describe how objects behave, as well as the tools for predicting this behaviour. Still, he fully admitted that he didn't know the origin of gravity. These laws appeared in print in 1687 in his *Philosophiae Naturalis Principia Mathematica* (commonly known as *Principia*), one of the greatest scientific books ever written. In Newton's own words :

> Every body perseveres in its state of rest, or uniform motion in a right line, unless it is compelled to change that state by forces impressed thereon.
>
> The alteration of motion is ever proportional to the motive force impressed (and inversely proportional to the mass of the body); and is made in the direction of the right line in which that force is impressed.
>
> To every action (force applied to a body) there is always opposed an equal reaction; or the mutual actions of two bodies upon each other are always equal, and directed to contrary parts.

Using the law of gravitation plus his laws of motion, Newton then rigorously deduced Kepler's three empirical laws of planetary motion. In other words, Kepler's laws were an empirical description of how the Universe seemed to work, and Newton's theory of gravitation now provided a physical explanation for them.

One minor caveat is that Kepler's third law is only approximate: the proportionality factor is, for the case say of a planet and the Sun, proportional to the sum of the two masses, so strictly speaking it will be different for each pair of bodies. However, in the case of the Sun and planets the variation is tiny since the mass of the Sun is far larger than that of any of the planets.

The first law addresses the motion of free bodies (free in the sense of not acted upon by forces), and is an improved restatement of Galileo's result, while the second states quantitatively how a body deviates from free motion and the third describes the effect experienced by a body when exerting a force on another object. Notice that uniform circular motion is now accelerated motion.

For our subsequent purposes in the book, and especially for our discussion of relativity, it is important to note that the first law implies that there is a privileged class of observers (called inertial observers) for whom free bodies do move at constant speeds in straight lines, or are at rest. In particular, Newton's mechanics is fully compatible with Galileo's relativity. For two inertial observers, the question of who is moving and who is at rest is unanswerable and meaningless, as no absolute velocities can be measured. The only thing that can be said is that they have a certain relative velocity.

As a short philosophical interlude, you may ask yourself whether the laws of motion are circular, or true by definition. A possible argument along those lines would be that these laws were historically expressed in terms of inertial coordinate systems, but these are defined as coordinate systems in which these laws are valid. The answer to this argument is that the significance of these laws does not lie in their truth—which is indeed, in this sense, trivial—but in their applicability. One could a priori conceive of and mathematically study any number of different types of coordinate systems, in which the mathematical expression of the laws of physics can have simpler or more complicated forms. The empirical fact that there exist countless systems of inertial coordinates is what makes the concept significant, since a priori we had no reason to expect the Universe to behave that way.

In fact, the above realisation is behind what is known as the principle of relativity, which asserts that for any material particle in any state of motion there exists an inertial coordinate system in terms of which the particle is (at least momentarily) at rest. This will be important when we discuss special relativity.

Moreover, two types of mass can be ascribed to each object:

> The inertial mass is the proportionality factor between the force applied to an object and the acceleration it acquires in response to it, in an inertial frame (in other words, it measures resistance to acceleration when a given force is applied).
>
> The gravitational mass measures the object's ability to attract (active mass) or be attracted by (passive mass) any other object.

In principle the gravitational active and passive masses could themselves be different, but in the case of Newtonian gravity their equality is ensured by Newton's third law: action always equals reaction. However, this need not be the case for other gravitational theories.

Newton's equivalence principle states that the inertial mass is proportional to the gravitational mass, but does not fix the proportionality factor. The reason for the equality will be understood when discussing Einstein's gravity, but from the perspective of classical gravitational theory it is simply a coincidence (or a brute fact of nature) with no fundamental explanation. Our purely conventional choice of units ensures that on Earth the two are numerically equal, which is why one is (within reason) allowed to mix together the concepts of mass and weight—as many students indeed do. However, if you travel to the Moon things are different: your mass will still be the same, but your weight will change because you will be in a different (in this case weaker) gravitational field.

Newton's equivalence principle then leads to the universality of free fall: the local acceleration of gravity is the same for all objects, independently of their mass or composition. A trivial consequence is then that all bodies must fall with the same acceleration in a gravitational field, which is sometimes anachronistically called Galileo's equivalence principle.

An interesting result found in the *Principia* is that the Earth and the other planets are oblate bodies, somewhat flattened at their poles and bulging at their equators. In the case of the Earth, the planet's surface is a few miles higher at its central belt than at the poles, hence the difference between the Earth's equatorial radius and polar radius, which are respectively 6378 and 6356 km. Thus the intensity of the Earth's gravitational field will not be the same everywhere, and as a result the Earth and the Moon pull on each other in a slightly off-centre direction. In turn, one consequence of this is that the orientation of the planet's axis of rotation slowly changes over time, like a spinning top. This effect is what astronomers call the precession of the equinoxes, and it was first noticed by Hipparchus. Newton's calculation showed that the period of this precession cycle is about 26,000 years, in excellent agreement with observations.

Notice that Newton's laws apply to all terrestrial as well as all celestial objects—this is the final demise of Aristotelian physics and its dichotomy. In the Newtonian view of the Universe, space and time are absolute. They are featureless objects, or a kind of arena where the play of Nature unfolds. They serve to define a universal and preferred reference frame—one may think of a universal network of equal measuring rods and perfectly synchronised clocks— and their nature is quite unaffected by the objects they host: in particular, any two observers will always agree upon the magnitude of a space or time interval.

Although everything in the previous paragraph might seem self-evident, the second part of it was later shown to be incorrect by Einstein, and many experiments provide confirmation for this. In any case it is hard to overestimate Newton's influence in the progress of science. His mechanics provided a sense of power over Nature that was hitherto unheard of. For the first time scientists could claim to understand at least a part of Nature's behaviour, and even though humanity was at the mercy of Nature's forces, these could be studied, understood, predicted, and sometimes tamed and harnessed for our convenience. Technological applications soon followed, and indeed Newtonian mechanics was one of the key catalysts of the industrial revolution.

While stating that all matter attracted all other matter through this gravitational force, Newton was careful to avoid materialism: he held that the force was not intrinsic to matter. This may be one of the reasons why Newton's ideas were immediately well received in Britain but criticised in continental Europe, especially by the many followers of Descartes. A few decades earlier Descartes had explained the motions of the heavens by entirely separating reality into two realms—matter and mind—and for Cartesians (such as Huygens) Newton was inappropriately mixing the two when he allowed this force of attraction to be imposed on matter.

Newton mathematicised the physical sciences. An important point of his mechanical philosophy is that everything must act by contact interactions. Newton's action at a distance implied mutual contact and propagation through a medium. Just as Aristotle did, Newton abhorred a vacuum, and therefore postulated an aether, a medium or agent by which many forces acting on matter throughout the Universe were propagated. The properties of such an aether would be the source of much experimental work and also great controversy in the following centuries: it had to be found everywhere, being less dense and more elastic than air, rarer within the dense bodies than in the spaces separating them, and as a result of all this, very difficult to detect experimentally. But unlike Aristotle, Newton did accept action at a distance. Note that gravity does operate, in our current view, via mutual contact through a medium and not a vacuum—but today that medium is a field.

4.3   Uranus and Neptune

Five planets can easily be seen with the naked eye, and so have been known from time immemorial. This situation changed dramatically on the night of 13 March 1781, when an object that soon proved to be a planet and is now called Uranus was discovered telescopically. The discoverer was William Herschel (1738–1822), a Hanover-born itinerant musician and amateur astronomer who by then had settled in England. Earlier on, at the age of 35, he had already decided to change careers and teach himself mathematical astronomy. This discovery was a serendipitous byproduct of an ongoing systematic sky survey of brighter stars. It took about half a year before it became evident that this was indeed a new planet rather than a comet.

The idea that our Solar System contained a whole other world, never before suspected, captivated the imagination of astronomers as well as that of the general public. An obvious question was whether Uranus—or indeed other possible planets—could have been observed previously but not recognised as such. Astronomers began going over the available star catalogues and records of earlier observers and found that Uranus, so named by the German astronomer Johann Bode (1747–1826), had in fact been seen at least 17 times prior to 1781, but always misidentified as an ordinary star.

The first known instance of such pre-discovery observations was in 1690 when John Flamsteed (1646–1719, the first Astronomer Royal of Great Britain) had listed it as '34 Tauri' in his star catalogue. However, the record—and also the bad luck award—goes to Pierre Lemonnier (1715–1799) who observed it and recorded its position in 6 out of 9 consecutive nights in January 1769 (as well as six other times), and still failed to identify it as a Solar System object. With the benefit of hindsight this would not have been difficult at all, given its motion relative to the background stars.

Although others were quick to speculate that this might be a planet and not a comet, for example Nevil Maskelyne (1732–1811), the orbit of Uranus was first calculated in August 1781 by Anders Johan Lexell (1740–1784) of Swedish origin, but the Professor of Astronomy in St. Petersburg, and independently by Pierre-Simon de Laplace (1749–1827). But as the observations accumulated and became more and more precise, they also began to reveal some puzzling discrepancies.

In 1821 the French astronomer Alexis Bouvard (1767–1843), then director of the Paris Observatory, compiled all the observations of Uranus then available and came across a major problem. Even after taking into account the gravitational influences of the giant planets Jupiter and Saturn in the orbit of Uranus, he could not reconcile all the data with what was predicted by Newton's laws of motion and gravitation. It should be noted that the differences between the calculated and observed positions were very small. If one imagines that the real and calculated positions of Uranus (which differed by only about 2 arcminutes) could have been placed side by side on the sky, they would have appeared as a single planet to everyone looking at them with the naked eye.

Going through the data in some detail, Bouvard further noticed that the problem could be ascribed to a discrepancy between the pre- and the post-discovery data, and cautiously suggested that the pre-discovery data might have errors larger than had been estimated. However, the data was coming from competent and mostly professional observers (Lemonnier's bad luck notwithstanding), and the errors that would be required to make the problem disappear were simply too large to be credible as an explanation.

George Airy (1801–1892, see Fig. 4.2) stressed these discrepancies in a review article that he published in 1832, and also discussed five possible solutions. The first three can be easily dismissed, but the last two are more plausible, as well as quite dramatic. It is worth considering them in some detail:

A catastrophic event, such as a collision with a comet, affected Uranus around the time of discovery and changed its orbit. This would save the calculations rather than the phenomena, but apart from the obvious issue of likelihood (why should the collision happen precisely around the time of discovery?), it was dismissed by the fact that the discrepancy between the calculated and observed orbits increased with time. Consequently, whatever was behind it was a steady long-term effect and not a 'one-off' occurrence.

Some kind of ether-like or cosmic fluid (which among others had been proposed by Descartes and, in a different framework, also by Newton) was affecting Uranus. But its existence had never been proved (in particular, it had no discernible effects on the motions of the other planets), and even if it did it could not account, at least by any known physical laws, for the observed perturbations of Uranus.

Uranus had an undiscovered, massive satellite. This would indeed perturb the orbit, but for any sensible object the orbital period (and therefore the characteristic time scale of the perturbations of the orbit) would occur on much shorter periods. In addition, the required mass and orbit of the satellite would make it easily visible with the telescopes available at the time.

Newton's laws were not exactly valid at these large distances—in particular the strength of gravity was slightly diminished at these distances. This hypothesis appealed to many astronomers at the time, and Airy himself preferred it.

There was a further, as yet undiscovered, planet in an orbit outside that of Uranus which perturbed its orbit. This solution had been defended by Bouvard, among others, since the late 1820s.

Indeed, in the early 1830s Bouvard formulated a plan for calculating the position of this hypothetical new planet and, once this had been estimated, scanning the appropriate region of the sky to search for it. This was the first known case of such a plan being spelt out in writing, but this was clearly understood to be a very difficult task. In broad terms, the idea was that, as Uranus approached Neptune, their mutual gravitational attraction would cause Uranus to be further ahead in its orbit than where it would be predicted to be if Neptune were not there. Conversely, as Uranus moved away from

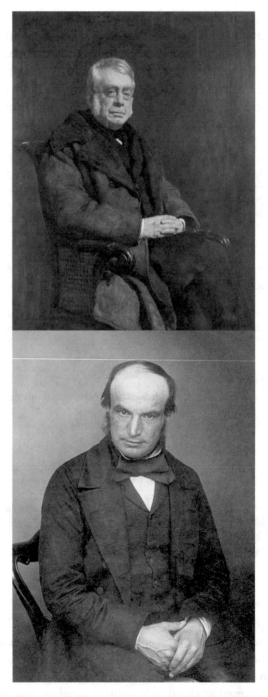

Fig. 4.2 George Biddell Airy and John Couch Adams (Public domain images)

Neptune, their mutual attraction would retard the position of Uranus in its orbit, and eventually Uranus would be back to where it would have been without Neptune. Therefore, Neptune's position in the sky should be towards the direction where Uranus was most ahead of its predicted track. Note that being in an outer orbit Neptune would be moving more slowly; on the other hand, one was implicitly assuming that the orbits of both planets were (at least approximately) in the same plane.

Several other analyses and plans followed in quick succession. Perhaps most noteworthy were those of the British amateur astronomer Thomas Hussey in 1834 (in a letter to Airy, which was met with a rather discouraging reply), but also Jean Valz (1787–1861) and Friedrich Nicolai (1793–1846). The last two both pointed out in 1835, independently, that the predictions for Halley's comet were also in some disagreement with observations. Valz also explicitly made the point that it would be remarkable to mathematically ascertain the existence of a planet that one could not yet observe.

By 1836 most astronomers had accepted the so-called exterior planet hypothesis, and not long afterwards this became widely known in the whole astronomical community (as opposed to being purely the privately voiced conviction of those specifically interested astronomers). A notable exception to this consensus was Airy himself, as he still preferred the modified gravity hypothesis and moreover thought that determining the new planet's orbit was unfeasible until much more data was gathered. In particular, in his reply to Hussey's letter of 1834 he had stated that the position in the sky of the new planet could not be determined until the nature of the irregularities was well determined from the data of successive revolutions of Uranus. Airy was by then the Astronomer Royal, and a practical man who frowned upon investigations that were too theoretical—which is somewhat remarkable and contrasts with the beginning of his career. It seems that he saw the modified gravity hypothesis as the least disagreeable. He was also rather sceptical of the abilities of young scientists (the way he treated his assistants was quite notorious).

Sometime in 1841, John Couch Adams (1819–1892, see Fig. 4.2), then a student in Cambridge, came across the problem of the motion of Uranus when he found Airy's 1832 report in a Cambridge bookshop, and almost immediately resolved to tackle the problem as soon as he had the time to do it. After graduating in Cambridge in 1843, Adams secured observational data on Uranus through James Challis (1803–1882), Plumian Professor of Astronomy and director of the Cambridge Observatory, a mile away—on 13 February 1844 Challis requested the latest available data on Uranus from Airy in order to pass it on to Adams. Having stayed on at Cambridge but busy with tutoring, Adams only had time for these calculations while on vacations at his parents' home in Cornwall.

Early in 1845 the director of the Paris Observatory, François Arago (1786–1853) suggested the same problem to Urbain Le Verrier (1811–1877, see Fig. 4.3), and he would mention the problem again on 22 September, this time with a more explicit request that Le Verrier investigate it. Earlier, in 1840, Arago had also suggested to Le Verrier a similar problem with the orbit of Mercury—we will return to this in the next section. Meanwhile, on 1 September 1845 Eugene Bouvard (nephew of Alexis) presented the *New Tables of Uranus* to the Paris Academy. These were substantially revised and improved, thanks in part to the high-quality data that Airy had been sending him. It's worth keeping in mind that by the 1840s Uranus had already been observed over about three quarters of its 84-year orbit.

In the middle of September 1845 Adams reported on his analysis to Challis during the summer, but given the sketchy nature of the indications, one can understand why Challis did not see these results as a compelling reason to interrupt his ongoing projects and start scanning the night sky to search for the new planet. Adams' solution gives what is called a dated mean motion, which is in fact a fictitious position used for the purpose of computational simplification. Further calculation is required to turn this mean motion into something that is useful for observational astronomers: moving from circular mean motion to the elliptical true or actual position. In other words, an astronomer unskilled in celestial mechanics calculations must know where in the sky he should point his telescope, and he might not know how to perform such calculations. That said, this was no excuse in the case of Challis, who was certainly mathematically skilled.

Perhaps being disappointed by the lack of action by Challis, Adams decided to pay Airy a visit in Greenwich on his way back to Cambridge from a vacation in Cornwall. Not being able to see him in person, Adams left a scrap of paper for Airy, with the orbital elements of the hypothetical planet. Airy in fact followed up this information with a letter to Adams: since he was still considering the hypothesis of the incorrectness of Newtonian mechanics and was sceptical of the new planet hypothesis, he asked for several small but reasonable clarifications, but for unknown reasons Adams did not respond. (It is known that he started to draft a reply, but never finished it or sent it.) Not having sufficiently explained his methods to Challis or Airy, he could not convince them to start an immediate search. It is worthy of note that Challis had England's most powerful refracting telescope, the Northumberland, an 11.75-inch refractor.

By this point Le Verrier was actively working on the problem, and on 10 November 1845 he published a detailed analysis of the motion of Uranus in a *Première mémoire sur la théorie d'Uranus* to the Paris Academy. This

Fig. 4.3   Urbain Le Verrier and Johann Galle (Public domain images)

identified several errors in the Bouvard analysis and calculated in more detail how Jupiter and Saturn were affecting the orbit of Uranus, but at the end of the day the problem persisted. There was an unacceptable discrepancy between the calculated and observed positions of Uranus. This publication reached England in December 1845, and Airy regarded it as a new and most important investigation.

A second and more complete analysis by Le Verrier was published in the 1 June 1846 issue of the journal of the French Academy of Sciences. He vehemently dismissed the four other hypothesis listed by Airy (though his arguments for the modified gravity dismissal were on somewhat thin ice), explicitly predicted that a trans-Uranian planet should exist, and gave details of its expected position in the sky. He also approached the Paris Observatory and suggested a search for the planet, but due to some earlier personal disagreements with his colleagues, he was not too popular there, and therefore after a brief search they lost interest.

In late June 1846 (possibly on the 23rd or 24th) Le Verrier's paper reached England, and as soon as Airy read it he realised he had seen a similar result the previous autumn—on the scrap of paper left at his home by Adams. Although Adams didn't specify detailed predictions for the position on the sky, there was no doubt that the so-called orbital elements of the two calculations were so close to each other that the predicted positions on the sky would correspondingly have to be quite close. In a letter of 28 June, which Airy received on 1 July, Le Verrier himself asked Airy to search for the planet on the sky. It was only at this point that Airy saw fit to suggest a search to Challis. Airy also computed the sky positions of the hypothetical planet for late summer and early fall. Interestingly, these calculations were based on the orbit predicted by Le Verrier rather than that of Adams. The former came with more details, so perhaps Airy thought it was more reliable, or perhaps he was simply relying on Le Verrier's seniority with respect to Adams (and therefore his greater experience with such calculations).

This brings us to the most fascinating—and also the most mysterious— point of this story. Airy was now the only person who not only knew that both Adams and Le Verrier were actively working on the problem, but also knew that they had obtained nearly identical solutions to it, and as far as is known he did not tell either of them about the other. Subsequent correspondence between them indicates that he certainly did not tell Le Verrier, although it is conceivable that he may have somehow warned Adams. He also did not inform Le Verrier of his plans to begin a search. It has been speculated that in the summer of 1846, Airy, Challis, and Adams conspired to ensure that Neptune would be discovered in Britain by keeping Adams' work unpublished, while they exploited the fact that the two calculations gave the same result to organise and carry out a search for the new planet.

Since the problem had been up in the air for a decade, why then was this only triggered by Le Verrier's results? Certainly, Airy's preference for the modified gravity hypothesis would have played a role—and being Astronomer Royal his opinions carried some weight both within and outside Britain. However, it should also be stressed that the various successively improved calculations by Adams yielded quite different predictions leading to results differing by more than 20 degrees on the sky—which would mean a very large area for any attempted search. Moreover, Adams' early calculations were in fact slightly simplistic. Thus on receiving Le Verrier's paper, Airy swiftly set in motion a huge search at the Cambridge Observatory, while suppressing news of Adams' calculations.

On 9 July, Airy wrote to Challis, asking him to begin a search for the planet with the Northumberland telescope, but when this and a follow-up letter reached Cambridge, Challis was away and so he only saw them on 18 July. Remarkably Airy didn't advise Challis to start by looking where Adams and Le Verrier said that the planet should be, or to try to identify the planet by searching for a disk. An obsessively methodical man, Airy instructed Challis to sweep over, three times at least, a zodiacal belt of thirty by ten degrees centred on the theoretical position of the planet, to complete one sweep before starting the next, and to map the positions of the stars therein. The point would be to identify the planet by its movement relative to the background stars, by comparing the positions in the sweeps, in much the same way that astronomers—and even secondary school students—search for nearby asteroids today.

Although from a methodological point of view this observational strategy was perfectly valid and thorough, and on its own would eventually lead to a detection of the planet if it was indeed there, it had one major drawback: it required a very long time. Thus one good reason for keeping a possibly long search secret would be to get a head start on any competitors. Airy himself estimated a total of 100 observing hours per sweep, and also recommended using a magnification of at least 120 (with the choices of eyepieces available to him, Challis decided to use 166), and also that the planet would appear no fainter than a star of ninth magnitude. Challis chose to map all stars up to magnitude 11. This would correspond to more than 3000 stars in the zone defined by Airy.

To put these instructions in context, this was the most extensive sky search in the history of astronomy up to that time. After delays due to other commitments and bad weather—the latter is a not uncommon occurrence in

Cambridge—Challis commenced his search on 29 July. He actually recorded the position of the object that would later be identified as Neptune on August 4, before pausing the search for a week because of moonlight and bad weather, and again when he resumed it on August 12. He decided to check out his methods by comparing the first 39 stars which he had recorded on 12 August against those that appeared on his 30 July records. If he had continued his comparison, he would have discovered that the new planet which he had recorded on 12 August had not been in the search area of 30 July. He had also recorded the planet on 4 August but he didn't look at that day's records immediately—by failing to do this, he missed the chance of making the discovery. Meanwhile, from July to September 1846, Adams provided Challis with various hypothetical planet positions, at heliocentric longitudes ranging from 336 to 315 degrees. The various calculations were not mere incremental refinements of one another, but actually used various different methods. This cannot have been reassuring for Challis.

British astronomy of this period was exceptional in having a large number of wealthy amateur astronomers who provided significant contributions in several fields. Late in August 1846 John Herschel (1792–1871), the son of Uranus' discoverer, visited William Dawes (1799–1868), a prominent amateur astronomer, and told him of the new planet hypothesis and the plans for its search. Unlike Airy, Dawes was quite impressed with the orbit calculations. Having only a small telescope he thought (wrongly, as it turned out) that it was not worth him trying to search for it, but he promptly wrote to his friend William Lassell (1799–1880), a brewer by trade, who at the time had just completed setting up the largest telescope in England, a 24-inch reflector, in his private observatory near Liverpool. Dawes sent Lassell Adams' positional data and urged him to look for a star with a disk in that region of the sky. When the letter arrived Lassell was confined to bed with a sprained ankle. He read the letter and gave it to his maid, who then promptly lost it. His ankle was sufficiently recovered on the next night and he looked in vain for the letter with the predicted position. His chance of fame had gone and, owing to the pressure of other work, he never undertook the search.

On 31 August, Le Verrier presented a third memoir on this subject to the Académie des Sciences. It was called *Sur la planète qui produit les anomalies observées dans le mouvement d'Uranus–Détermination de sa masse, de son orbite, et de sa position actuelle*, and as the title suggests the memoir left nothing to be desired in the way of completeness. In fact, it explicitly said that the planet should be found about 5 degrees east of the star δ Capricornii. He also pointed out that the predicted angular size, of 3.3 arcseconds, was large enough to produce a distinguishable disk, and boldly suggested that proper motion would not be necessary to identify it: a simple inspection of physical appearance should be enough.

At the same time, Adams was carrying out yet another revision of his own calculations, which he summarised in a letter to Airy on 2 September. By then, it no longer mattered: the new calculations were too late to influence any searches for the planet. At the same time John Hind (1823–1895), an expert on the discovery of asteroids—he would discover 10 in the following decade—who worked at George Bishop's Regent's Park Observatory and had previously worked as an assistant at the Royal Greenwich Observatory, received a copy of Le Verrier's memoir and, knowing that Challis was searching for the planet, advised him to take notice of the fact that the new planet should be visible as a disk—which as we will see Challis eventually did, although too late.

To reiterate an earlier point, one interesting aspect of the calculations by Adams is that he always described his planet as some kind of mathematical abstraction: a set of numbers and coordinates (which astronomers call 'elements') which needed to be balanced in the proper way. Le Verrier, in contrast, had previous experience in calculating orbits of asteroids and comets and, more to the point, in interacting with observers. He clearly envisaged a real physical object, and he was always much more thorough in publishing specific positions in the sky for his hypothetical planet.

What an observer needs in order to know where to point the telescope is the prediction for the so-called true heliocentric longitude at a given time, that is the position of the planet in the zodiac, as it would appear if it were seen from the Sun. But this is difficult to calculate directly, mainly due to the ellipticity of the orbit. Therefore, as a first step the calculations were done with a mean heliocentric longitude, the position on a notional circular path with uniform motion. The effect of the ellipticity can then be subsequently added as a correction. The difference between the two calculations was that Le Verrier was careful always to indicate the true heliocentric longitude (or even both, as in the case of the letter to Galle we shall discuss presently), while Adams only did so towards the end, if at all—for example, in the data he gave Airy he only used the mean heliocentric longitude. This difference was crucial: the true and mean heliocentric longitudes (and hence the orbital positions on a given date) can diverge markedly, especially for orbits of significant ellipticity—as was in fact being assumed by both.

As the summer of 1846 drew to an end, Le Verrier was becoming increasingly dismayed with the inertia of French astronomers in searching for the planet. On 18 September he decided to write to Johann Galle (1812–1910, see Fig. 4.3), an assistant at the Berlin Observatory. He apologised for not having replied for a year to an enquiry about Mercury that Galle had made, and he took the opportunity to ask weather he could, in his words, "take a few moments" to examine a region "where there could be a new planet to discover". He also explained that its diameter of around 3 arcseconds would be "readily distinguishable in a good telescope". Galle received the letter on 23 September, and immediately approached the observatory's director, Johann Encke (1791–1865), to ask for permission to search for the planet. After a few attempts during that day, Encke reluctantly agreed. In the course of the last of the conversations between the two they were interrupted by Heinrich d'Arrest (1822–1875), a young student-astronomer, who had overheard Galle's request and asked to be allowed to participate in the search.

That same night Galle and d'Arrest opened the dome over the 9-inch Fraunhofer refractor, the largest then available at the Berlin Observatory. Galle pointed the telescope at the exact position predicted by Le Verrier but could not find any obvious disk. d'Arrest suggested using a star map. Going through a pile of disordered charts, d'Arrest found a new detailed map of that area of the sky, which had been printed in 1845 but had not yet been distributed to other observatories. Galle then returned to the telescope while d'Arrest sat at a desk and began to check the stars on the map as Galle called out their positions and appearances. Only a few had been checked when Galle described a star of eighth magnitude, and d'Arrest immediately exclaimed "That star is not on the map!" The eighth planet had been found, in less than half an hour of observing and within a degree of the position predicted by Le Verrier.

As observatory director Encke was rapidly called over to confirm the discovery, but by the time he arrived the planet was too close to the horizon for observation. When they observed it again the following night, it had moved slightly—the unmistakable sign that it was a Solar System object rather than a star. Galle wrote immediately to Le Verrier, "The planet of which you indicated the existence really exists!" The name that was eventually agreed for the planet, Neptune, was Le Verrier's suggestion (Galle had suggested Janus), although he later changed his mind and wanted it named after himself.

On 16 September John Hind had written to Challis. It was clear he had not heard of the calculations by Adams, but he had read Le Verrier's third memoir and had said he would look for Le Verrier's planet. He also mentioned that his colleague the French astronomer Hervé Faye (1814–1902) intended to do the same, and Faye is quoted by Hind as expecting to spot the planet as a two arcsecond disk, following Le Verrier's advice of 31 August. Starting on 21 September Challis continued the search in the area that Le Verrier had suggested, although he only acknowledged the shift of strategy towards searching for a disk on 29 September, that is, after Galle's discovery. Indeed, Challis's misfortunes were not yet over.

On 29 September the news of the discovery had not yet reached England, but this was the day when Airy (who was traveling in Europe and was then at Gotha) heard of the discovery. On this same day Challis received a copy of Le Verrier's third memoir, emphasising that the planet should be identifiable by its disk, with the obvious implication that a scan was not necessary. This finally convinced him to disregard Airy's instructions and pay more attention to the physical appearance of the celestial objects he was cataloguing. On that same night he scanned Le Verrier's region, and just as he was interrupting the observations in order to attend a dinner he hesitated over one of the 300 or so stars he had checked, noting in his log book "last one seems to have a disc", with 'last one' subsequently crossed out.

During the dinner with William Kingsley (then chaplain of Sidney Sussex College) in Trinity College, Challis mentioned his observations and Kingsley asked the obvious question: would it not be worthwhile to look at it again with a higher magnification? Challis replied "Yes, if you will come up with me when dinner is over we will have a look at it." The sky was clear when they walked from Trinity College to Challis's lodgings at the university observatory. They had planned to go right up to the dome, but Mrs. Challis insisted on giving them tea before they began their observations. This certainly was one of the bitterest cups of tea in the history of astronomy: by the time they had finished their tea the sky had clouded over and observations were impossible, and the same happened the following night. Again Challis omitted to follow up a valuable clue and so missed the great opportunity of his life.

As it turned out, Challis wasn't even the first person to identify the planet in Britain. John Hind observed it on the night of 30 September, having received a letter from Brunnard in Berlin that morning which contained its coordinates, and he immediately wrote to a number of people, including Challis and Adams, and also The Times' editor. On 1 October a letter in The Times by Hind announced Galle's discovery with the headline *Le Verrier's Planet Found*, and also said that the planet had been observed in London on the previous night. Thus Challis first heard of the discovery of Neptune by reading The Times. To add insult to injury, the letter also said that the planet's disk was quite noticeable—even though the largest telescope that Hind had available was a 7-inch one, so quite a bit smaller than the Northumberland.

Upon hearing of the discovery, on 30 September, Challis decided to go through his observations since 29 July. He quickly noticed that he had recorded its position on 4 and 12 August, that it wasn't on the list of observations of 30 July (which was of the same zone as 12 August), and that the object whose disk he had suspected on 29 September was indeed the new planet. Challis was mortified, and wrote to Airy "after 4 days of observing the planet was in my grasp, if only I had examined or mapped the observations". Through his own lack of initiative and his submission to Airy's intellectual dominance, Challis had lost his chance of a great discovery, and lost the respect of his colleagues.

With hindsight, it was soon noticed that there were also a number of previous sightings (and recorded positions) of Neptune, starting with Galileo on 28 December 1612 and 27–28 January 1613, while he was observing Jupiter to study the motions of its satellites. It is interesting to speculate how astronomy might have evolved if Galileo had discovered Neptune. He was somewhat unlucky because, at the time he observed it, it was at a stationary point on its orbit (switching from direct to retrograde motion) and therefore moving very little with respect to the background stars. Other pre-discovery observations were due to Jérôme Lalande (1732–1807) on 8 and 10 May 1795, John Herschel on 14 July 1830, and John Lamont (1805–1879), a Scottish born astronomer who lived most of his life in Munich, and who recorded its position on 25 October 1845, and on 7 and 11 September 1846—only days before the actual discovery!

Just after the discovery, while Airy was returning to England, he stopped at Altona on about 5 October to see Heinrich Christian Schumacher (1780–1850), who was the editor of *Astronomische Nachrichten*. Here he carefully read the extensive manuscript that Le Verrier had by then sent to the journal (and which would appear in print between 12 and 22 October) explaining the mathematics of his discovery—this much Airy admitted in a letter to Le Verrier on 14 October. This was over a month before Adams published any of his calculations, and it is therefore quite possible that Adams would have received information on the details of Le Verrier's work from Airy before writing up his own work for publication.

Despite the strong controversy in England and France over priority claims once the work of Adams became known in late 1846 (which not only involved the astronomical societies and newspapers of both countries but even reached the highest levels of politics and diplomacy), on 30 November 1846 the Royal Society of London, which had known of Adams's work for several weeks, awarded its highest scientific honour, the Copley medal, to Le Verrier for his brilliant discovery of the planet. Both Challis and Airy strove to give the impression that they hardly knew Adams, while simultaneously giving him credit for one of the greatest astronomical discoveries ever. Why would they want to do that? Challis had spent six weeks not finding a planet that two Germans (one of whom was a young student) found in less than half an hour with an inferior telescope. And Airy's own reputation was at stake because as Astronomer Royal he had set up and effectively directed Challis' search.

The British claims were untenable for several reasons. First there was no publication by Adams of his methods or results until more than seven weeks after the discovery in Berlin. His pre-discovery work was effectively secret, and this was almost certainly done deliberately. What Adams did do within weeks of the discovery was to correctly determine not only the distance of the new planet but even the correct position of its node (or equivalently, its semi-major axis) and orbital inclination—in other words, he determined its true orbit. For this he made use of the positions that Challis had recorded in August during his sky search, which of course no other observatory had. So these observations did prove important for the orbit determination—as they did for the forthcoming claim of precedence. Adams was the first to calculate the true orbit of Neptune from observations, and quickly released these for publication (in a letter to Challis on 15 October, which Challis communicated to the Athenaeum on 17 October)—weeks before he made known his predicted pre-discovery elements, supposedly from the year before.

Airy's politically inspired behaviour is difficult to judge, but may be explained in the following way. His first reaction to Adams (in 1845) was a mixture of caution and curiosity, though perhaps not quite enthusiasm. Hearing of Le Verrier's paper of 10 November 1845, he may perhaps have warned Adams of the competition. Then, seeing Le Verrier's paper of 1 June 1846 he launched (on 29 June) a secret and thorough sky search to try to ensure that Neptune would be found in Cambridge, being aware that although the search would surely find the planet if it did exist, it could well take some time. Hearing of Galle's discovery of 23 September 1846, he realised that the race had been lost. He continued to regard Le Verrier as the prime discoverer, while perhaps helping Adams to publish his results. By December 1846 Airy had become the public villain who ignored the public hero Adams in 1845 and so a mixture of pressure, fear, and opportunity made him slide back to the Cambridge party line.

By the end of 1846 Neptune had been observed by most astronomers in Europe and America. As soon as he read The Times headline Herschel wrote to Lassell saying "Look for satellites with all possible expedition". Lassell began observing on 2 October, and on 10 October recorded a first sighting of what proved to be Neptune's largest moon, now called Triton. Other astronomers confirmed Lassell's discovery the following July.

This mathematical discovery of a new planet provided one of the great triumphs of Newtonian mechanics, but this story is also interesting due to some other lessons. Adams was certainly a pioneer, as was Le Verrier, in applying mathematical techniques (now known as perturbation theory) to planetary motions, both real and hypothetical. Although how much Adams actually achieved before seeing Le Verrier's calculations in print (or hearing about them from Airy) is up for discussion, when looking at the predictions he subsequently made, the very least that can be said is that he was on the right track. The fact that Le Verrier's results were more consistent with each other may simply have been due to his already significant previous experience with such calculations, while Adams was doing them for the first time and thus learning them as he went along.

On the other hand, Adams completely failed to communicate his results either to his close colleagues (who had the means to start an observational search much sooner than they actually did) or to the rest of the scientific community. A scientific discovery does not simply consist of asking a question and finding an answer, be it by calculation, experiment, or observation. The wider context matters: one must understand the relevance of the question and answer to be able to infer whether or not the result is what was expected, and therefore what is a genuinely new result—a discovery. And the result, within the appropriate evidence and context, must also be communicated to the rest of the scientific community (and eventually also the general public) who can then scrutinise it and decide whether or not it is scientifically valid and how important it is. In short, a discovery has both a public and a private side.

In this case Adams only completed part of this task. On the other hand, Le Verrier succeeded both in predicting the planet's position and in convincing several others—in France, Britain, and Germany—to search for it. Thus the credit for the mathematical discovery of Neptune is Le Verrier's alone.

4.4   Mercury and Vulcan

While the issue of the discrepancies in the orbit of Uranus was being hotly debated during the 1830s, another superficially similar issue also arose. It was noticed that Mercury's orbit also deviated from what was predicted by Newtonian mechanics. In Mercury's case, this was related to the fact that the direction of its perihelion (the point in the orbit closest to the Sun) advances by a small amount each orbit. This effect exits for all planets, and indeed it is predicted by Newtonian mechanics, but in the case of Mercury, not to the extent that was observed.

Seen from Earth, Mercury's perihelion precesses by 5600 arcseconds per century, of which 5026 are due to the precession of Earth's axis of rotation, leaving 574 arcseconds to be explained due to the other planets. However, Newtonian theory only accounts for 532 of these: specifically 278 are due to Venus, 154 to Jupiter, 90 to the Earth, and 10 to the other planets. This leaves a discrepancy of 43 arcseconds per century. This is an extremely tiny amount. It would take four thousand years for the discrepancy to mount up to, say, the apparent angular size of the Moon, and three million years for it to amount to a complete turn about Mercury's orbit. Nevertheless, if one trusts both the Newtonian calculations and the available observational data, such a difference requires an explanation: if theory and data do not match, there must be something wrong with at least one of them.

Incidentally, this effect also exists for other planets but it's much smaller since they are further away from the Sun: for Venus it's 8.6 arcseconds per century, while for the Earth it's only 3.8 arcseconds per century. An additional difference is that Mercury's fairly eccentric orbit makes it much easier to detect the perihelion shift than is the case for the orbits of Venus and the Earth, which are much closer to being circular.

In 1846, following the discovery of Neptune (and in particular the difficulty in agreeing on a name for it), Le Verrier and Jacques Babinet (1794–1872) decided on a preemptive strike and suggested that a hypothetical intra-Mercurial planet might be responsible for the discrepancies, and immediately gave it the name Vulcan. In the 1850s Le Verrier tackled the problem in a systematic way, and came up with gradually more refined predictions for its orbit.

As in the case of Uranus, several alternative explanations were suggested and explored. Let us briefly consider some of them. An interesting possibility would be a change in the value of the mass of Venus. A planet's mass can be easily calculated if it has satellites moving around it: all one has to do is use Kepler's laws. Unfortunately, Venus has no satellites. Its mass could only be determined in other more indirect ways, and if it actually was 10% larger than the astronomers of the mid-nineteenth century had estimated, that would suffice to account for Mercury's motion. But that explanation has a fatal weakness: that extra mass would also affect the orbit the Earth, in a way that is not actually observed, so Le Verrier himself rapidly eliminated the Venus solution.

Other possible solutions that would eventually be suggested were an intra-Mercurial asteroid belt as an alternative to a single inner planet, a Mercury satellite (recall that there was an analogous proposal in the case of Uranus), or the flattening of the Sun—in much the same way as the Earth is oblate, as shown by Newton. Interestingly, and again as in the case of Uranus, there were also proposals suggesting that Newtonian gravity didn't exactly apply. One such possibility was a correction to Newton's law of gravitation depending on the inverse of the cube of the distance (in addition to the usual inverse square law), since one can prove mathematically that such a small additional term would produce a precessing ellipse. On the other hand, Simon Newcomb (1835–1909), whose calculations actually established the correct 43 arcsecond discrepancy (Le Verrier's calculations had originally found 38 arcseconds) suggested that an empirical correction to Newton's law with an inverse power of 2.0000001574 would solve the problem, but this seemingly tiny correction turned out to be ruled out by lunar orbit calculations.

Observing a planet inside the orbit of Mercury would be extremely difficult, as you may realise if you have ever tried to observe Mercury. You can see it with the naked eye, but only for a total of a few weeks every year. For a planet even closer to the Sun, this might not be possible at all in ordinary circumstances. Alternatively, you may use a telescope, but this is not necessarily much better. The telescope must be pointed very close to the Sun, so the sky will necessarily be bright. If the telescope optics is not perfectly aligned this light is likely to give you false images, and of course there is the risk of accidentally pointing to the Sun, with undesirable consequences for both the telescope and you.

A good time for the observations would be during total solar eclipses, since for a few minutes you will have a dark sky, but these eclipses are of course rare. The other option is to wait for the equally rare occasions when the planet passes in front of the Sun's disk, which astronomers call a transit. You will then see a small, round dark spot moving across the Sun. The size of this spot depends on the planet's size and distance: in the case of Venus the disk is substantial and unmistakable, but in the case of Mercury—see Fig. 4.4—it's not uncommon to see larger sunspots.

Transits of Venus are very rare—if you missed the ones in 2004 and 2012, the next one will be in 2117. Those of Mercury are somewhat more common, happening at irregular intervals roughly 13 or 14 times per century. The transits happen only in May or November (with November transits the more common in the ratio of 7 to 3) and at successive intervals of thirteen, seven, ten, and three years. For a hypothetical Vulcan orbiting closer to the Sun than Mercury, and assuming that its orbit is approximately in the same plane as the Earth's, one would expect the transits to be more frequent.

In December 1859 Le Verrier, whose work on Vulcan was well known in France, received a letter from a physician and amateur astronomer called Edmond Modeste Lescarbault (1814–1894), who claimed to have observed a transit of the hypothetical planet in his small observatory earlier in the year, on 26 March. Le Verrier visited him unannounced and upon questioning him was satisfied with his observational skills and hence with his claim. For example, Lescarbault said that he believed that what he had observed could not have been a sunspot since it was moving faster than such a thing would have done,

Fig. 4.4   Transit of Mercury on 11 November 2019 (Credit: NASA/Joel Kowsky)

and that having observed a transit of Mercury in 1845, he assumed that he was also observing a transit this time around.

Thus on 2 January 1860 Le Verrier formally announced the discovery of Vulcan to a meeting of the Académie des Sciences in Paris. According to Le Verrier's calculations Vulcan's period of revolution was 19 days and 17 h and the mass only one seventeenth of Mercury's mass. As it happens this would have been too small to account for the deviations of Mercury's orbit, but one could reasonably make the argument that perhaps this was one of the members of that intra-Mercurial asteroid belt. Later that year there was a total eclipse of the Sun, but despite many efforts to find Vulcan, nobody did.

Conversely, both that year and throughout the subsequent two decades there were several dozens of claims of sightings of the hypothetical planet, either as it was transiting the Sun or during total solar eclipses. Some of these were due to competent and experienced astronomers, but none of the sightings could ever be corroborated. With each new report Le Verrier tinkered with Vulcan's orbital parameters, often using them to predict and announce dates of forthcoming Vulcan transits—and when these failed to be observed, he again revised his calculations. Most false alarms could have been due to sunspots, or to asteroids or comets transiting the Sun, or faint stars near the Sun during eclipses, while a few were never clearly explained.

When Le Verrier died, in 1877, he was still convinced that he had discovered another planet, but without the main champion for the idea it was slowly abandoned by most astronomers as an interesting but insoluble problem. In 1900 Edward Charles Pickering (1846–1919) announced that as a result of his photographic survey having failed to detect an intra-Mercurial object, there could be none brighter than the fourth magnitude, and by 1909 William Wallace Campbell (1862–1938) had announced a more stringent analysis which showed that there was nothing inside Mercury's orbit that was brighter than the eighth magnitude. That magnitude limit roughly corresponded, with the assumption of a rocky body with a typical composition and density, to a body with a diameter no larger than 48 km. It would take at least a million such bodies to account for the missing 43 arcseconds per century in the movement of Mercury's perihelion.

The answer suddenly and unexpectedly came in 1915, when Einstein calculated the correction to the Newtonian value predicted by his general theory of relativity. Since the gravitational field of the Sun at the distance of Mercury's orbit is, at least comparatively speaking, fairly weak, general relativity only predicted a fairly small correction, which turned out to be exactly the missing 43 arcseconds per century. Einstein himself, in a letter to Fokker, recounts that calculating this correction in his theory and finding a result of 43 arcseconds caused him palpitations of the heart for several days. His friend and biographer Abraham Pais further states that this was the strongest emotional experience of Einstein's scientific career, and possibly of his whole life.

With Einstein's result most astronomers abandoned the search for Vulcan. Of course, nothing precludes the existence of previously unknown comets or small asteroids in the vicinity of the Sun, but their dynamical effects will be tiny and they will have no significant impact on Mercury's orbit. Today, some astronomers are still looking for the so-called Vulcanoid asteroids, which might exist in the region where Vulcan was once sought. None have been found so far, and searches have ruled out any such asteroids larger than about 60 km.

Comparing the stories of Neptune and Vulcan, the important lesson is that two similarly-looking scientific puzzles (a planet's observed orbit disagreeing with the calculated expectations of a given theoretical framework) were initially addressed in much the same way (as can be seen by the similarities between the proposed explanations in the two cases), but turned out to have opposite explanations. In the first case the theory was correct (and was indeed further vindicated by its remarkable confirmed prediction of the solution), while in the second it wasn't and the true solution turned out to be changing the theoretical framework.

This also shows that scientists, being human beings like everyone else, are not immune to prejudices, and not infrequently it is not easy to keep an open mind for alternatives when one standard explanation seems preferred. In science one never runs the risk of being too radical—the risk is always not being radical enough. One has an innate tendency to attack new problems with old methods and techniques that have been used over and over again and have previously worked well, but this provides no guarantee of success since one may be facing something completely different.

4.5   The Physics of Gravity and Satellites

Why do astronauts float when they are inside the International Space Station (ISS)? Most people would intuitively say that it's because there is no gravity in space, or because gravity is very weak there, but that's actually the wrong answer.

If I'm floating inside the ISS and drop an apple in front of my nose, it will keep floating in front of my nose. So relative to me it does not move, which is an important observation. Now let's suppose that I go to the top of the leaning tower of Pisa, and do a modified version of the experiment that Galileo never did: I take my apple with me, and just when I jump from the top of the tower I put the apple in front of my nose and drop it. For simplicity let's assume an ideal situation without atmosphere, and therefore without friction. Then the universality of free fall predicts that the apple and I will fall at the same rate, so relative to me the apple still does not move (we shall not discuss what happens when we both reach the ground). There must therefore be something common to both situations, since my apple experiment produces the same result in both cases.

To understand what is common, suppose that I take a cannon to the top of the Pisa tower. I also take two cannonballs. The cannon shoots one of them with some horizontal speed, while I drop the other one from rest, from the same height and at the same time. In the ideal case without atmosphere they still reach the ground at the same time, and in principle this will be the case regardless of the speed of the cannon. In practice this does not happen because at some point we will need to take into account the fact that the Earth is not flat but spherical. In other words, for large enough initial speeds the distance travelled by the cannonball will be such that the assumption of a locally flat Earth will no longer be reasonable.

It is surprising that very few people have a good intuition for the amount of curvature of the Earth. The best way to approach it is to think that when you go to the top of a mountain (or even a tall building) and look far away, the Earth's surface at that distant point will be some distance below where it would have been if the Earth was flat. Ignoring effects of the local topography (mountains, valleys, and so on), you can use this distance as a measure of the Earth's curvature. That number is worth knowing: it is approximately 5 m for each 8 km. So when you look at a point 8 km away (which you can very easily do, at least on a clear day), the Earth's surface there is on average 5 m below the plane perpendicular to your feet.

Why is this number useful? Because 5 m is how much an object falls in 1 s, if it is dropped from rest near the Earth's surface. Suppose that my cannon is powerful enough to provide a speed of 8 km/s. Then after 1 s the cannonball has fallen 5 m and travelled 8 km. But 8 km away from the cannon the Earth's surface is also 5 m below where it was at the point of departure, so the cannonball is still at the same height above the ground. So I no longer have a cannonball—what I have is a satellite.

What this shows is that on Earth 8 km/s is the orbital speed: in other words, the horizontal speed I need to give an object near the Earth's surface to put it in orbit around it. So the common point between my two apple experiments— and the reason why the results are the same in both cases—is that we're falling near the Earth or, to use a technically more appropriate term, in free fall. In the case of the Pisa tower we just fall for a few seconds (again, we shall not discuss what happens when we hit the ground), while on the ISS we're constantly falling because the ground is also falling—at exactly the same rate as the ISS. As Douglas Adams succinctly put it:

> There is an art to flying, or rather a knack. The knack lies in learning how to throw yourself at the ground and miss.
>
> Douglas Adams (1952–2001)

This is what has to be done to put a satellite in space. Note that we have obtained this 8 km/s orbital speed on essentially geometrical grounds, and seemingly without using any gravitational properties. (But if you think a bit more carefully you will see that actually gravity is lurking in there ...) A horizontal speed of 8 km/s will lead to a circular orbit, while a larger or non-horizontal speed will lead to elliptical orbits.

In an ideal situation, without friction, one can have satellites orbiting at any chosen altitude. In practice, for a planet like Earth with a substantial atmosphere and therefore with a significant amount of frictional drag, it is not viable to have satellites orbiting for any significant amounts of time at altitudes below about 180 km.

Also notice that 8 km/s will be the orbital speed for satellites near the Earth, but this speed will decrease for satellites orbiting further away—as expected from the fact that the gravitational field will become weaker. For example, in the case of the distance of the Moon this speed is only about 1 km/s.

Another commonly used concept is the escape velocity, which is the speed that must be given, on the surface of a particular body, to make another object escape the gravitational field of that body. In the case of the Earth this has the value of about 11 km/s, but it is not independent of the orbital speed. Anyone with a knowledge of high school physics should be able to show that the escape velocity always exceeds the orbital speed by a factor of $\sqrt{2}$.

This orbital speed is the difficult part of getting into orbit—by difficult, I mean costly in terms of energy (that is, fuel), as we will see shortly. Gaining the required altitude is the easy part. This is why, if you have ever seen a rocket being launched, in the first few seconds it moves straight up (after all, you don't want to accidentally hit anything near the ground), but after that its trajectory will become fairly close to horizontal, so that it can start gaining that required speed.

Artificial satellites can be put in space for a number of reasons. Among those currently operational about 41% are commercial, 17% are military, 18% are governmental, and 24% are scientific or have multiple functions. Depending on these functions they will have to be placed at different altitudes, as we will now discuss. Broadly speaking there are three classes, with self-explanatory names: low Earth orbit (LEO), medium Earth orbit (MEO), and high Earth orbit (HEO). Broadly speaking, and again for currently operational satellites, 63% are in LEO, 6% are in MEO, 29% are in HEO, and the rest are in significantly elliptical orbits—one example of the latter is the so-called Molniya orbit. Both these sets of numbers are likely to change significantly in the next few years.

Incidentally, you can easily see the ISS (see Fig. 4.5) and quite a lot of other satellites with the naked eye or at most with binoculars, if you know when and where to look. This is normally in the hours after sunset or before sunrise, so that the satellite is still illuminated by the Sun while for you on

Fig. 4.5 The International Space Station on 23 May 2010, photographed from the Space Shuttle Atlantis (Credit: NASA/Crew of STS-132)

the ground the Sun has already set or not yet risen. Several internet sites, for example Heavens-Above, will determine and provide lists of visible passes for any specified location. Satellites visible to the naked eye are typically on the higher end of LEO.

Most LEO satellites orbit at altitudes of 200 to 300 km. The defining feature of such an orbit is its orbital period of about 90 min. Incidentally, at an altitude of 300 km the gravitational field is 9.37 m per second per second as opposed to the surface value of 9.81 m per second squared, which shows that the former answer that there is no gravity in space is literally wrong.

In some circumstances you can see two consecutive passes of the ISS: if you see it passing, wait slightly more than one hour and a half (as its altitude is slightly higher, about 400 km), and see it passing again, in that time it went all the way around the planet. This also shows that if you have the necessary technology to build and launch an intercontinental missile you can reach any target in 45 min or less.

The canonical example of satellites at this altitude are spy satellites, other examples being those to monitor natural catastrophes (including global warming) and agriculture. The need for LEO stems from the wave nature of light. When using an optical system—in other words, a telescope or camera—the size of the smallest objects that can be clearly seen (what astronomers and physicists call the resolution) is given by

$$R = \frac{h \times \lambda}{D},$$ (4.1)

where in our present scenario h is the altitude of the satellite observing the ground, λ is the wavelength of light used for the observation, and D is the diameter of the primary component (lens or mirror) of the telescope. Taking as an example the Hubble Space Telescope (HST), which has a primary mirror of 2.4 m as we will further discuss shortly, if we use it to look at the Earth's surface from HEO its resolution will be 8 m, while if it observes from LEO its resolution will be 6 cm. The former is of little military use, while the latter is extremely useful.

However, there is also a shortcoming associated with LEO: observation times will be very short. If you want to photograph, say, your house, you have about 1 min per day. Just like when you see the ISS passing overhead, you only see it for a couple of minutes (after all, it is moving at about 8 km/s), anyone on the ISS itself can only see you for a similar amount of time. One orbit and approximately 90 min later the ISS will be approximately back to the same point in space as compared to the centre of the Earth, but you won't be there because the Earth is rotating—depending on your latitude, you might be more then 2000 km away. The next near passage will be 12 h later, and it will likely be at night if the first was during the day. What this means is that in order to have a network of satellites capable of observing any point on the planet at short notice one would need many thousands of satellites. So nowadays most of the work previously done by satellites has been taken over by drones—being at lower altitudes, say 20 to 30 km, even smaller telescopes will give an adequate resolution.

The canonical example of MEO satellites are those of the Global Positioning System (GPS), and those of analogous systems from the European Union, Russia, and China. These constitute what are called constellations of satellites, whose orbits are in various different orbital planes chosen to ensure continuous coverage of the whole surface of the planet. In the specific case of the GPS the constellation has six orbital planes with four satellites each. The orbits are circular, with altitudes and speeds respectively of approximately 26,600 km and 14,000 km/h, and periods of half a sidereal day. There are also some further satellites in the so-called parking orbits—they are not parked in the usual sense of the word (they are of course moving as they orbit the Earth), but they are in space and ready to replace any malfunctioning satellite at fairly short notice. Having to launch a replacement satellite would clearly take longer.

The constellation is built such that at least six are visible at any moment from any point of the planet—which allows for some redundancy, since only four are needed to fully determine your position on the planet. Actually, only three would be needed if you had a clock as accurate as those on the satellites themselves, but these are rubidium or caesium atomic clocks, and therefore you will only have one of those if you are in a good physics laboratory. They emit in the frequencies of 1226.60 and 1575.42 MHz and are passive systems, meaning that the system does not require any information to be provided by the user. One reason for this is that the GPS is in part a military system: it has a high-resolution military version whose signal is encrypted and a lower-resolution open access version. The choice of MEO is the result of a simple trade-off: at lower altitudes more satellites would be needed for the reasons we have already discussed, while at higher altitudes the communications would require more powerful emitters and also more sensitive receivers.

In its open-access civilian version, the GPS has an accuracy of 10 m. This translates to an accuracy in the clocks involved of 30 nanoseconds—which is just the time it takes light to travel those 10 m. It turns out that at this level of clock accuracy there are relevant effects coming from both special and general relativity that must be accounted for—in other words, a Newtonian GPS would simply not work to any useful level of precision. We will revisit this point in the next chapter after we have discussed relativity, but for the moment bear in mind that the GPS may well be the only aspect of your everyday life for which relativity is needed—for the rest, Newtonian physics is quite sufficient.

The canonical HEO satellites are in geosynchronous orbits, which have altitudes of ca. 35,786 km, periods of one sidereal day (that is, 23h 56min 4s), and orbiting speeds of 3 km/s. There is one special orbit among these: the one above the equator. This is called the geostationary orbit, because it is the only possible orbit for which an observer on the ground will always see the satellite in the same position in the sky: in other words, it doesn't move with respect to the ground, because its period is exactly the same as that of the Earth's rotation around its axis.

It is this geostationary orbit that is used for communications and meteorological satellites. If you have a satellite dish at home, you will have noticed that it is always pointing in the same direction, and this is only possible if the satellite whose information it is receiving is not moving in the sky. If you are at mid-latitudes in the Northern hemisphere you may also notice that the dish is approximately pointing due south, precisely because the satellite that your dish is pointing to is over the equator. (However, it doesn't have to be exactly south, since the satellite need not be at exactly the same longitude as you are.)

The reason for having meteorological satellites at this altitude is also clear. In order to observe the atmosphere and its evolution you don't need a resolution of centimetres—on the contrary, you want to be able to look at a large patch of the atmosphere, in order to track, say, a hurricane developing over the Atlantic. Moreover, if you have built a satellite to study the weather in Europe, you don't want the satellite wasting half of its time on the other side of the planet—so a geostationary orbit is exactly what you want. Notice that there is only one geostationary orbit, and space out there, though vast, is certainly finite. As more and more satellites are placed there, several issues arise, having to do with allocations of this space and possible interferences between neighbouring satellites.

Among the telescopes in space there are of course scientific (astronomical) telescopes. These were theorised by Lyman Spitzer (1914–1997) in 1946, and there is actually a telescope currently in space named after him. The first space telescope was OAO-2, which was launched in 1968 and stayed operational until 1973, at an altitude of about 770 km. Actually this was a set of 11 small ultraviolet telescopes, the largest of which had an aperture of 41 cm.

Putting astronomical telescopes in space has clear advantages. There is no atmosphere or light pollution, you get a better image quality for the same telescope size, and you are also able to observe at frequencies that are blocked by the Earth's atmosphere, especially the ultraviolet, X-rays, and gamma rays, all of which are (as we have gradually found out) astronomically interesting. On the other hand, the disadvantages are also obvious: the costs will be higher, maintenance is much harder (actually, in most cases, impossible), and all of its components must be robust enough to survive the rather violent launch process and still function when they have reached the intended orbit.

Astronomical space telescopes can be broadly classified into two types. Surveyors have a pre-defined task of mapping the full sky (or at least a significant fraction thereof) for a single or a small number of scientific goals—even if that data can almost always be used for a lot of additional science—while observatories carry a much broader science program that is constantly being selected from proposals made by the astronomical community. That said, some telescopes have a dual role, with a survey part that is pre-defined and a fraction of the observation time open for observatory-type proposals.

The most important and best-known recent space observatory is undoubtedly the HST, a NASA/ESA telescope launched in 1990 and still active. Through its various generations of instruments, it can observe in the ultraviolet, visible, and near infrared, spanning the wavelength range from 0.115

to 1.7 microns. The telescope is in LEO, at an altitude of about 540 km; its speed is about 7.6 km/s and it orbital period 95.4 min. The primary mirror has a size of 2.4 m, and the telescope dimensions are about 13.3 by 4.3 m; at launch, its weight was about 11,110 kilograms. This is therefore a rather large satellite. If you live relatively close to the equator you can sometimes see the HST passing overhead.

An example of a space surveyor was the Planck satellite, which was operational from 2009 to 2013. This carried out a full-sky survey in the far infrared and microwave wavelengths (from 0.3 to 11.1 mm, corresponding to frequencies between 27 GHz and 1 THz). Its main scientific goal was to study the anisotropies of the cosmic microwave background, which are an important cosmological probe as we will see later. This was a much smaller satellite, with approximate dimensions of 4.2 by 4.2 m, and a weight of 1950 kilograms. It was placed at the second Lagrange point of the Sun–Earth system, at a distance of 1.5 million kilometres.

These Lagrange points are physically interesting. They are the points in a system of two orbiting bodies where a third smaller one can maintain its position with respect to them (everywhere else, the third body will necessarily orbit one of the two main ones). There are always 5 such points, in the plane of the orbits: three of them are along the line joining the two bodies and are always unstable, while the other two form equilateral triangles with them and are stable provided the ratio of the masses of the two main bodies is greater than 24.96. The first three were actually discovered by Leonhard Euler (1707–1783), while Joseph-Louis Lagrange (1736–1813) discovered the missing two later.

Several planets have satellites in their Lagrange points with the Sun. These are generically called Trojan satellites; Jupiter, for example, has more than a million of them. Placing artificial satellites at the Lagrange points, either of the Earth–Sun or the Earth–Moon system as the case may be, is often very convenient. In the case of the Earth–Sun system, the point L1 is useful for satellites aiming to study the Sun, since the Earth will never be in the way, while L2 is convenient for satellites wanting always to have the Sun behind them (to prevent it from spoiling the observations). This is the case of the Planck satellite and many other scientific satellites. For the Earth–Sun system, L1 and L2 are 1.5 million kilometres from the Earth; this is about 1% of the distance to the Sun and about 3.9 times the distance to the Moon.

Nowadays, we put satellites in space using rockets. In physics terms this is actually a terribly inefficient way of doing it, but it is the best we have found so far. The reason why rockets are inefficient is that they fly by pushing down the fuel they burn (as opposed to, say, an airplane, which flies by pushing air down). Moreover, rockets need to carry both their fuel and their oxidant, while an airplane only needs to carry the fuel—the oxidant is the oxygen in the air. This means that for every kilogram of payload to be put in space one will need between 25 and 50 kilograms of fuel. The exact number will vary with the type of orbit that one is aiming to reach: the higher the orbit, the more fuel will be needed. This shows that the efficiency of rockets can be no higher than about 4%. Launching satellites into orbit using airplanes (at least as a first step), whenever possible, is far more efficient.

One way to mitigate this limitation is to build rockets with multiple stages, as opposed to monolithic ones. Once each stage has burned its fuel, it can be discarded. Even if the orbital speed has not yet been reached (the goal, remember, is to reach 8 km/s), the system will have some fraction of this speed and will now be lighter and therefore easier to accelerate further. The Space Shuttle (which unfortunately no longer flies) is a good example. The spacecraft itself had a weight which depended on the mission but was around 68 metric tons, while the full system at the moment of launch had a weight of about 1931 metric tons. But 125 s after launch, when the solid rocket boosters were discarded, the system's weight was only 30% of the initial value—the other 70% was solid fuel that had already been burned. Thus putting in space 1 kilogram of anything costs about 20,000 euro, and that is basically the cost of the fuel. You can then try to scale this up for trips to the ISS, the Moon, or Mars.

Let's think of the case of sending a spacecraft to Mars. You may ask what is the fastest way to travel from the Earth to Mars, in the sense that the amount of time taken by the journey is minimised. Or you may ask what is the shortest way to get to Mars, in the sense that the distance you cover is the smallest possible. Or you may ask what is the cheapest way to get there, in the sense that your fuel bill is the lowest. The answer to the three questions will be different, and unless there is a compelling reason to do otherwise (say, an astronaut stranded on the surface of Mars), the high cost of getting to space means that there is an obvious incentive to use the third option. But what is the most energy-efficient way of doing it? The solution is called a Hohmann transfer orbit, see Fig. 4.6. Interestingly, this was found in 1925 (long before the space age), which illustrates the point that it is a conceptual physics problem rather than a practical (or engineering) one. In these orbits fuel is only spent at the beginning and end of the transfer between the two bodies. In the transfer itself one simply lets gravity and Kepler's laws do their work, at zero cost in fuel.

The clever realisation leading to the Hohmann solution is that an object on the Earth (or launched from it) is already orbiting the Sun. Kepler's laws

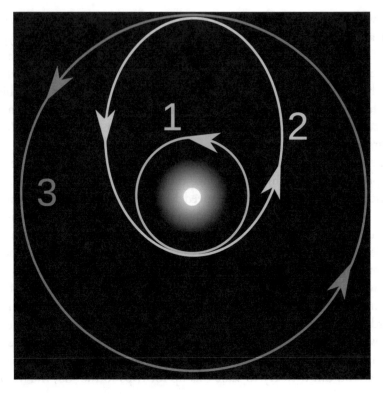

Fig. 4.6 A Hohmann transfer orbit from the Earth (1) to Mars (3). Note that the sizes of the two orbits are not to scale (Public domain image)

show that the distance to the apoastron (the most distant point in an orbit) can be increased by increasing the spacecraft speed at the periastron (the nearest point in the orbit), while the distance to the periastron can be reduced by reducing the speed at the apoastron. Thus in order to send a spacecraft to Mars one just needs to apply an appropriate tangential force that gives it an initial velocity increase in the direction of the Earth's motion. Than the spacecraft will be in an elliptic orbit around the Sun whose periastron coincided with the orbit of the Earth and the apoastron with that of Mars, and will travel to Mars without spending any fuel. Once it reaches Mars it needs another force from retro-rockets to decelerate and move back to a circular orbit, but now one that coincides with that of Mars—effectively being captured by it. It's only these two velocity kicks that cost fuel, everything else is free, courtesy of gravity. Indeed, engineers in the field typically do not talk about forces or accelerations, but simply talk of delta-v's.

Of course, this can only work if the spacecraft and Mars reach the same point of the Mars orbit at the same time, so this transfer orbit is not always available. Given the distances and orbital periods of the Earth and Mars—which again are related through Kepler's laws—there is what is called a launch window every 25 months. If you have wondered why ESA, NASA, and other space agencies only send missions to Mars every two years or so, this is the reason: they need to save on their fuel bills. If instead one wants to send a spacecraft to Venus the principle is the same: one needs a transfer orbit which makes it fall to an inner orbit, and in that case the initial velocity kick will be in the direction opposite the Earth's motion. In this case a launch window will be available every 19 months.

This 8 km/s orbital speed has other interesting consequences, too. One of them is that it corresponds to a very high kinetic energy. This is a problem if you want to make a satellite or spacecraft return to the ground, for example, because it is carrying astronauts, or samples collected from other Solar System bodies. This kinetic energy has to be dissipated by heating the atmosphere, because the alternative is not good: if it heats the spacecraft, it will get so hot that it will disintegrate. This is what happened to the Space Shuttle Columbia in 2003. One consequence of this large kinetic energy is that the angle of re-entry into the atmosphere needs to be carefully chosen: too small and you will bounce back to space, too large and you will disintegrate.

This also raises the issue of the cost–benefit relation of crewed space missions by comparison to those carried out by robots. Clearly, at the moment that relation is not favourable. The cost of making a spacecraft habitable by humans is very significant (it typically increases the cost by a factor of at least ten), and being an astronaut is a risky job—more than 3% of those who have orbited the Earth have died either in space missions or training for them. (Additionally, there will of course be some long-term health effects.) And if one looks closely, the impact of scientific experiments done for example at the ISS is remarkably low, and on its own would clearly not justify its costs. Naturally, there are many other reasons to keep the ISS functional, as there were many reasons, beyond scientific exploration, to go to the Moon, but this is a discussion that will certainly be revisited in the coming years.

Putting satellites in space also comes with the risk of space junk, from decommissioned or malfunctioning satellites, final stage rockets, protection shields, and various types of debris. There are currently more than 100 space launches per year, and as of early 2020, more than 2300 operational satellites (and this number may be much larger soon), but many more are still in orbit. However, the problem is not just the big satellites. The kinetic energy corresponding to the 8 km/s orbital speed is so large that a 1 cm screw colliding with a satellite or the ISS can cause very significant damage. All these objects must therefore be carefully monitored. This is easy to do for objects larger than about 10 cm (estimated to be more than 22,300), but as we have already mentioned, even smaller fragments than that are dangerous. It has been estimated that there should be, in the Earth's neighbourhood, more than one million objects larger than 1 cm and more than 170 million objects with sizes larger than 1 mm.

Of course, objects in LEO (where the overwhelming majority of the space junk is) still feel the effect of the atmosphere, and if nothing else is done to prevent it they will slowly lose altitude and eventually fall back to Earth, most likely being burnt up in the atmosphere. But this process is very slow: for a typical sized satellite at an altitude of 600 km, the re-entry time would be 25 years or more.

Collisions of small space junk fragments with other satellites (whether or not they are functional) are therefore a major problem. Although when a satellite stops working one doesn't necessarily know the reason why it happened, it has been estimated that more than 30 satellites have been damaged by collisions. But the problem is broader than that, and is generally known as the Kessler syndrome. Since collisions occur at high speed, each collision will create a large number of new fragments; more junk means a higher probability of subsequent collisions, which mean more junk, and so on. Nowadays, significant collisions typically occur every few years.

Space agencies like ESA and NASA resort to collision avoidance manoeuvres several times per year. The orbits of these debris fragments are always known with some uncertainty, so typically one can only calculate the probability that a collision will occur, say with the ISS. If that probability is greater than 1 in 10,000, then the ISS will be moved to a different orbit until the debris passes through. Naturally this can only be done if the satellite that one wants to move has a propulsion system that has some leftover fuel and can be controlled. Technology for cleaning up space debris exists in principle, but is limited by international treaties: a space system that can destroy your space junk can also destroy someone else's operational satellites.

4.6   Asteroids and Comets

Not all objects in the neighbourhood of the Earth are artificial satellites or space junk that we have sent there. There is a very large number of asteroids, and a smaller number of comets, whose orbits pass quite near—or even intersect—the Earth's orbit, and these can occasionally collide with the Earth.

It is worth considering four of those relatively recent collisions, each of which is interesting and/or representative in a different way:

> If the collision happens at a grazing angle, the object can bounce on the Earth's atmosphere and go back into space, just as you can bounce a small flat stone on the surface of water. An example of this happened on 15 August 1972 over Utah, with a small asteroid with an estimated diameter of 3 m, whose collision happened at a minimum height of 57 km and speed of 15 km/s. Incidentally, this object has been captured on video, which you can find on YouTube.
>
> Perhaps the most famous such collision, at least in recent times, happened in Tunguska (Siberia) on 30 June 1908. The object had an estimated diameter of 45–80 m and exploded at an altitude of 5–10 km; the uncertainties in both of these numbers stem in part from the fact that it has not been clearly established whether the object was an asteroid or a comet. A collision of this magnitude typically happens once per millennium.
>
> A more recent (and also more frequent) example is that of Cheliabynsk on 15 February 2013. This had an estimated diameter of 18 m and a mass of ten thousand metric tons, and disintegrated at an altitude of 23 km. An event of this magnitude typically happens once per century.
>
> The first object that was detected in space and identified as being on course for a collision with the Earth (with a warning time greater than zero) disintegrated over Sudan on 7 October 2008. It was identified about 20 h before the collision, and had an estimated diameter of 4.6 m and a mass of 80 metric tons, with the disintegration happening at an altitude of 37 km. Less than 10 kg of meteorites were recovered, in 280 different fragments, from the Sudan desert.

Again their speeds of 8 km/s—but often significantly higher—imply a large kinetic energy, and it is this kinetic energy that causes the damage. In the case of Chelyabinsk there is no record of deaths, but the shock wave due to the energy of the disintegration caused injuries to an estimated 1600 people on the ground, as well as a significant amount of material damage.

Incidentally, since the Sudan event, there have only been two other objects whose collisions with the Earth were predicted before they actually happened. The second case is from 2014 (a 3 m object, with a warning time of 21 h) and the third is from 2018 (an object with an estimated size between 2.5 and 4 m, with a warning time of 8 h).

These objects are generically known as Near Earth Objects (NEOs) and just as in the case of space debris, there is an obvious incentive to know where they are. There is no universally accepted definition of a NEO, but a typical one is any object (whether it's an asteroid or comet) whose orbit has a minimum distance to the Sun less than 1.3 times the size of Earth's orbit. It has been estimated that there are more than 1000 NEOs with sizes larger than 1 km, and more than one million with sizes larger than 30 m. Various projects are working toward the goal of finding them. For example, Project SpaceGuard aims to catalog at least 90% of the potentially dangerous objects with sizes larger than 140 m, but it is thought that less than 40% are currently known. On the other hand, more than 90% of objects with sizes of 1 km or larger are known, and there is no known threat from them in the next 100 years.

What happens if a collision occurs will depend on the size and composition of the object. Broadly speaking, if it's smaller than about 30 m it will disintegrate in the atmosphere, and damage on the ground may be limited. For sizes between 30 and 100 m an impact is still unlikely, but the shock wave will certainly cause extensive damage. Objects with sizes larger than 100 m will reach the ground, and their high kinetic energy will create a crater whose size may be 10 to 20 times larger than the size of the object itself.

The process of disintegration in the atmosphere is what physicists call an airburst. The object's kinetic energy is rapidly transformed into thermal energy and kinetic energy of air while the object fragments, vaporises, and decelerates. They typically occur for objects with diameters between 10 and 100 m. The simpler term 'explosion' is often used, but it is not quite correct: although there is a rapid energy release, no chemical or nuclear explosion is taking place.

For an asteroid, the initial speed as it approaches the Earth's atmosphere is typically around 17 km/s, but this is gradually lost by ablation. A long-period comet has a typical speed that may well be three times higher, but only about twice as much kinetic energy, since its density will typically be lower, being more icy than rocky or metallic. Roughly speaking, objects with sizes of 1, 10, and 30 m will disintegrate at altitudes of 50, 30, and 20 km, with speeds of 16, 13, and 9 km/s, respectively. For an object that reaches the ground, the typical speed at the moment of impact is only around 5 km/s.

For long-period comets, the estimated frequency of collisions with the Earth is one in every 43 million years. For asteroids, estimated frequencies for various sizes are 1 per year for 4 m objects, 1 every 10 years for 10 m objects, and 1 every 300 years for 30 m objects; less than 1% of these 30 m objects are thought to be known. Incidentally, small fragments of such collisions occasionally hit humans directly—the estimated frequency is 1 human impact every 9 years, while about 16 buildings are hit every year.

Our society has warning scales for various physical phenomena, and there is also one for collisions with asteroids or comets. This is known as the Torino scale, and was introduced in 1999. It has 11 different levels, from 0 to 10, and at the time of writing all known objects are in level zero, though this can change as new objects are discovered and previously known objects are upgraded or downgraded when their orbits are further studied. What level on the scale is assigned to each object depends on two parameters. One is obviously the probability of impact: as new objects are found, their orbits will not be precisely known, so all we can do is calculate a probability of impact given the currently available data. But the second parameter is not the radius or the mass of the object—it is its kinetic energy, and it should already by obvious to you why physicists have made that choice.

If one such object is heading our way, what can we do? Ideally, we would like to change its orbit by a sufficient amount to ensure that the collision is avoided. The difficulty is again the very high kinetic energy (and also momentum) of the asteroid or comet. A conventional impact of a satellite with a typical size and speed can only change the asteroid's speed by something of the order of 1 cm/s. This means that if the asteroid was previously moving at say 15 km/s, after the collision it will be moving at 15.00001 km/s. This is a tiny change, but it does accumulate: over a period of 10 years it adds up to about the Earth's diameter, meaning that if beforehand the Earth and the asteroid would have been at the same place at the same time (thus colliding), this will no longer be the case after the collision with the satellite. So the limiting factor ends up being time: with a 10 year warning it is possible to do this, at least for objects with sizes up to about 500 m whose orbits and compositions we know reasonably well, but with a 20 h warning it is not.

The possibility of using nuclear explosions (on the asteroid or near it) is often discussed, but it has its drawbacks. In the short term an explosion near the surface would remove some material by ablation, with a fusion bomb being more efficient than a fission bomb. In the long term an explosion inside the asteroid that caused it to fragment might work, though a central explosion will not change the centre of mass trajectory, and the damage caused by multiple fragments with similar trajectories may in some cases be greater than the damage caused by a single fragment with the same mass, so at best these would be last-resort (or last-minute) solutions.

A more intelligent solution is what is called a gravitational tractor. Instead of throwing the satellite against the asteroid, one leaves it in its vicinity, and waits for its gravitational field to slowly deflect it. Apart from the obvious advantage of not destroying the satellite, this will work in principle for an asteroid or comet of any mass. No knowledge of their composition is needed, and one can guide the satellite to provide real-time adjustments as the orbit is studied (by the satellite itself) and therefore becomes better known. The limiting factor is again time—for any object that is sufficiently large for it to be essential to avoid the collision, a gravitational tractor will generally require several decades.

To end on a brighter note, we can briefly consider shooting stars. On a typical day, about 100 metric tons of meteoroids, interstellar dust, and space junk intercept the Earth atmosphere, the overwhelming majority of which disintegrates; of this, a ton or two (on average) are satellites, rocket components, or other artificial junk. You may wonder whether the Earth is thereby gaining mass, but the answer is no: some of the Earth's atmosphere escapes into space at a comparable rate. But the number of 100 tons can increase by a factor of ten at the time of meteor showers.

Most of this material burns up in the atmosphere, but particularly heavy or dense fragments can reach the ground and potentially cause some damage. The only recorded victims of space junk are five sailors aboard a Japanese ship who were injured in 1969, and Lottie Williams, a woman from Oklahoma who was hit on the shoulder in 1997 by a 10 by 13 cm piece of metal which was subsequently identified as being part of the propellant tank of a Delta II rocket that had been launched the year before.

When you see one of these meteors—colloquially called a shooting star— you are not seeing the object itself, which is typically the size of a grain of sand. What you are seeing, once again, is the effect of the high kinetic energy associated with the 8 km/s orbital speed. This energy heats the atmosphere and causes a track of incandescence, starting at an altitude of about 90 km above the ground and almost always ending by an altitude of 80 km, and it is this incandescence that you see. Most of the shooting stars that you see from the ground will be less than 200 km from you.

On a normal night (without a meteor shower) and with good observing conditions (no Moon or light pollution), you can see an average of about 6 meteors per hour, usually called sporadics. During a meteor shower you can see many more. The relevant number is what astronomers call the zenithal hourly rate, which is the number you would see under ideal conditions, including the apparent direction of origin of the meteors—called the radiant—being directly above your head. Thus the typical number you will actually see can be quite a lot lower. Numerically, the zenithal hourly rate is quite similar to the number of meteoroids in a cube of side 1000 km. Thus the cloud of dust that the Earth is going through actually has a very low density by everyday standards. An important point is that you can see more shooting stars in the second half of the night, which is when the effects of the rotation and translation of the Earth add and the meteoroids hit the Earth head-on.

The trajectories of the meteoroids from a given cloud are approximately parallel in space, but when we see them from the ground going through the atmosphere they seem to come from a specific point, the aforementioned radiant. This is due to the Earth being in motion through the cloud, and it is analogous to what you see when you walk or drive through rain or snow. Even though they are falling vertically, the faster you move, the more the rain or snow seem to be coming at you.

5

The Physics of Relativity

In this chapter we discuss Albert Einstein's special and general theories of relativity. We start by a brief look at their origins, and then discuss in more detail the overall physical content and a few specific properties of each of them. Part of the difficulty in understanding relativity when one first encounters it is that some of its properties seem strange to our classical intuition, but we also discuss some of the many tests that have been done over the last 100 years and support their validity.

5.1 The Decline of Classical Physics

One could say that when it comes to physics, the nineteenth century was the century of electromagnetism. While Newtonian mechanics was robustly established and indeed saw major triumphs such as the discovery of Neptune which we discussed in the previous chapter, intense experimental and then theoretical activity was under way to establish the properties of electrical and magnetic interactions and obtain a suitable physical model for them.

Starting in the late 1850s, James Clerk Maxwell (1831–1879, see Fig. 5.1) developed a mechanical model for a common description of electric and magnetic forces—in more modern terms he unified the two interactions, showing that they were simply two different manifestations of a single underlying force. This was first described in his 1865 article *A dynamical theory of the electromagnetic field*, and later and in greater detail in his book *A treatise on electricity and magnetism* (1873). This model describes it as a Newtonian-type

© Springer Nature Switzerland AG 2020
C. Martins, *The Universe Today*, Astronomers' Universe,
https://doi.org/10.1007/978-3-030-49632-6_5

Fig. 5.1 The young James Clerk Maxwell at Trinity College, Cambridge. The date of the photo is unknown (Public domain image)

mechanical interaction between various parts of an as yet unidentified ethereal substance.

One of the properties emerging from this description is that in an otherwise empty space the perturbations of this substance—called electromagnetic waves—always travel at the speed of light. This therefore offered an explanation for what light actually is: transverse waves (meaning that the direction of oscillation is perpendicular to that of propagation) of the same medium that was behind the electric and magnetic phenomena. At the time the value of the speed of light had been known for about two centuries: in 1675 Ole Roemer (1644–1710) had used Jupiter's satellites to measure it (see Fig. 5.2) and obtained a value that was only about 20% too low. More importantly, he showed that it was finite.

But in the prevailing view, waves had to be transmitted through a material medium, so the fact that we see light from the Sun and distant stars means

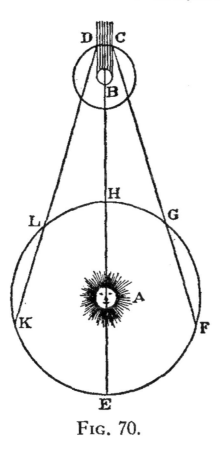

FIG. 70.

Fig. 5.2 Illustration from the 1676 article on Roemer's measurement of the speed of light. He compared the duration of Io's orbits around Jupiter (C–D and B, respectively) as the Earth moved towards Jupiter and away from it (F to G and L to K, respectively) (Public domain image)

that the space between them and us should be permeated by this medium. As we have already discussed, Descartes, like Aristotle, believed that all space was a plenum, and there could exist no vacuum: an aether permeated all space, and instantaneous action at a distance was senseless. Newton's laws actually contain such an 'action at a distance', since no speed of propagation enters the equations. But Newton's description of the gravitational interaction is also based on transmission of action by direct contact through an intermediary.

This is one example of an emerging trend in the latter half of the nineteenth century. Despite the multiple successes of Newtonian physics, by the end of the century, a number of experimental results that were difficult to explain within its framework had accumulated, suggesting the possible need for novel theories. It is worth listing the most notable among these:

Although light was described as a wave propagating in a medium, the properties of this medium could not be consistently described, and the very existence of the medium could not be confirmed experimentally.

The dynamical equations for electromagnetism obtained by Maxwell were quickly seen to be inconsistent with those of Newtonian mechanics.

As we have already discussed in the previous chapter, the observed orbit of the planet Mercury disagreed with that predicted by Newtonian physics by an amount that was too large to be ignored.

Netwonian physics also failed to predict the correct behaviour of materials at very low temperatures.

Newtonian physics predicts that a body at a stable and constant temperature should have infinite energy.

It turned out that explaining all of these inconsistencies required entirely new theories, which were developed in the early twentieth century and changed our view of the Universe. The first two of these motivated the development of the special theory of relativity, while we already know that the third one could only be explained by the general theory of relativity. Finally, explaining away the last two inconsistencies required the development of quantum mechanics.

For our present purposes, the most interesting and revealing of these inconsistencies has to do with the relation between Newtonian theory and the theory of electromagnetism. What can now be called Galilean (or Newtonian) relativity describes the behaviour of physical objects in different frames moving with respect to each other at some relative speed, and the laws of mechanics transform appropriately (and are preserved) between any two of these frames. However, the same Galilean relativity acts differently on the laws of electromagnetism. Specifically, Maxwell's equations in one frame would not transform into Maxwell's equations in another frame if one used the Galilean transformations.

To give a simple example, if you are on a train and throw a ball with a horizontal speed of 1 metre per second, while the train is itself moving with a speed of 2 metres per second in the same direction, then an observer sitting on the train platform and watching the train go by would see the ball moving with a speed of 3 metres per second. On the other hand, if you throw the ball vertically instead of horizontally, both observers will agree on the ball's vertical speed. If you instead carry a flashlight and throw photons, Maxwell's equations say that they should move at the speed of light, both for you and for an observer sitting on the train platform. And yet Galilean relativity would predict that if you see them moving at the speed of light inside the train, for the observer on the train platform the speed of the photons should be the speed of light plus 2 metres per second.

Was there a reason why inertial frames preserved the laws of mechanics but not those of electromagnetism? That would be strange—why would different parts of physics behave differently at such a fundamental level? On the other hand, could Maxwell's equations be wrong? One might ask what kind of transformation between frames would be needed to preserve Maxwell's equations, in the same way that the Galilean transformations preserve the laws of Newtonian mechanics. These were obtained by Hendrik Lorentz (1853–1928) in 1892 and are now known as Lorentz transformations (we will return to them shortly), but this was a mathematical solution to a problem whose physical interpretation was far from clear.

Another difference, which we have already mentioned above, is that Maxwell's laws of electromagnetism contain an explicit speed (the speed of light) which is therefore the speed of propagation of the electromagnetic interaction. On the other hand, in Newton's laws there is no explicit speed. This means that formally the speed of propagation of the gravitational interaction should be infinite. If the Sun suddenly disappeared (for example, because Vogons destroyed it to make way for a hyperspace bypass), then the Earth should instantly feel the absence of its gravitational field. Again, this was somewhat puzzling: why should two different interactions, seemingly described by the same type of mechanical model, behave in such dramatically different ways?

Galileo's principle of relativity implies that no mechanical experiment can determine an absolute constant uniform velocity, and this principle is maintained in Newtonian physics. Now Maxwell's equations appeared to enable the possibility of defining an absolute reference frame: although they do not specify with respect to what that velocity should be measured, it presumably had to be the speed relative to some particular frame. In other words, the frame where the aether was at rest would be the one for which the speed of light had the standard value—today this value is defined to be exactly 299,792,458 metres per second.

There are two possible conclusions to be drawn from this. One possibility is that absolute velocities can be determined using experiments involving light, which would be a third difference between the two theories: while Newtonian mechanics could not single out any special inertial frame, Maxwell's seemed to do so. The alternative would be to accept that light must move with the same speed in all reference frames as Maxwell's equations predict. But this is plainly incompatible with Newtonian mechanics, which predicts that the speed of light should depend on the speed of its source.

Meanwhile, on the experimental side the situation was equally puzzling. If the aether was indeed the medium in which light propagated, one should be able to infer its properties and ultimately detect it experimentally, but as has already been mentioned the former could not be done consistently and attempts to do the latter came out empty-handed.

Since we manifestly receive light from distant stars the aether had to permeate interstellar space, and in particular the Earth had to be moving through this aether as it orbited the Sun. The aether had therefore to be rather tenuous: if it caused any significant amount of friction it would surely have affected the Earth's orbit in a noticeable way. But the fact that the Earth was moving through this aether implied that an aether wind should be detectable, in the same way that you feel the wind hitting your hand when you stick it out of the window of a moving car or train.

Assume that light travels in air and in glass with speeds c_a and c_g respectively, as measured with respect to the aether. Now let's measure the ratio of the two speeds assuming that the Earth is moving with respect to the aether at some speed v. If our apparatus for measuring the speeds of light in air and glass happens to be placed such that the light is moving in the direction of the Earth's motion with respect to the aether, Newtonian physics would predict that we should measure a ratio

$$v = \frac{c_a + v}{c_g + v}, \tag{5.1}$$

while if we rotate the apparatus by $180°$ such that the light is now moving in the opposite direction we should expect to find a different value,

$$v = \frac{c_a - v}{c_g - v}. \tag{5.2}$$

And yet, as was first shown by François Arago (1786–1853), one always finds the same value, regardless of how the experimental apparatus is oriented.

In order to explain this Augustin-Jean Fresnel (1788–1827) suggested that transparent substances somehow trapped the aether and dragged it along so that its speed was locally zero. On the other hand, the phenomenon of stellar aberration, discovered by James Bradley (1693–1762) as early as 1727, implied that the air did not trap aether at all. In fact, Thomas Young (1773–1829), one of the founders of the wave theory of light, had said back in 1804:

> Upon consideration of the phenomena of the aberration of the stars I am disposed to believe that the luminiferous aether pervades the substance of all material bodies with little or no resistance, as freely perhaps as the wind passes through a grove of trees.
>
> Thomas Young (1773–1829)

So the aether seemingly goes through all objects without much resistance, but nevertheless some of it gets trapped by transparent substances. A key test of this explanation was performed by Hippolyte Fizeau (1819–1896). He sent light through tubes containing water flowing in different directions. Under the expectation that the water would drag along some aether, the measured speed of light should change. Fizeau did detect a change (in agreement with Fresnel's hypothesis), though a much smaller one than would have been expected. So on the one hand, astronomical observations implied that the Earth did not drag the aether, while the Fizeau experiment suggested that transparent media did drag at least some of it. Could it be the case that the aether behaved differently in the heavens and on Earth?

The final and ultimately better known attempt to detect the motion of the Earth became known as the Michelson–Morley experiment. This was actually a succession of gradually improved and more sensitive experiments by Albert Michelson (1852–1931) and Edward Morley (1838–1923). A light beam is split into two and the two new beams are sent in perpendicular directions for some distance; they are then reflected back and recombined. Depending on the distances travelled by each beam, the combined one will produce some pattern of light and dark fringes. But now rotate the experimental apparatus. Since the speeds of the two beams with respect to the aether will change, so will the times they take in their trips before they recombine, and consequently the observed interference pattern will change, too. Or at least it should have changed, except that despite repeating the experiment in multiple settings (from a basement room to the top of a mountain) the experimental result was always the same: the rotation of the experimental apparatus did not change the interference pattern. Therefore, no motion through the aether could be detected. Clearly, there had to be a problem with the aether: the Michelson–Morley experiment showed that it must be dragged along by the air, while stellar observations showed the exact opposite.

One last attempt to explain the negative result of the Michelson–Morley experiment was due to Lorentz and (independently) George Fitzgerald (1851–1901), who postulated that matter moving through the aether is compressed by just the right amount to lead to that negative result. In other words, the aether does change the speed of light but additionally it also compresses all moving objects, and somehow the two contributions together lead to no effect in all experiments. Lorentz showed that for this to be the case an object with a speed v relative to the aether should be contracted by a factor

$$\sqrt{1 - \frac{v^2}{c^2}} . \tag{5.3}$$

The reciprocal of this quantity is usually denoted γ and known as the Lorentz factor. Lorentz also suggested that this could be due to a molecular force, whose behaviour in different inertial frames had therefore to be similar to that of the electromagnetic forces.

To summarise, the inferred properties of the alleged aether fluid were clearly inconsistent, and Newton's mechanics and Maxwell's electromagnetism were also mutually incompatible. The aether had to be a very tenuous medium and yet have the ability to compress all materials (even the strongest materials) so as to hide itself from experimental detection. It seemed that Nature was conspiring to prevent the detection of this privileged reference frame at absolute rest.

This in fact led Henri Poincaré (1854–1912) to postulate that only relative velocities between observers could be detected, not absolute velocities—a statement sometimes known as the postulate of relativity. Notice that this questioned the existence of the aether, and in fact came remarkably close to special relativity, as we will see in a moment.

It was in this context that Albert Einstein (1879–1955, see Fig. 5.3) began his seminal contributions to physics. In 1905, which has been dubbed his 'miraculous year', Einstein published five papers, each of which would by itself have been enough to assure him a place in the history of physics. Four of these papers were published in volume 17 of the Annalen der Physik—making it a highly prized collector's item. The fifth paper was effectively his doctoral thesis.

Einstein's quantitative studies of Brownian motion, discussed in two of the papers, allowed Jean Baptiste Perrin (1870–1942) to experimentally demonstrate the existence of atoms, which was still a debatable issue at the time. Indeed, before and in between these two works you could be a respected

Fig. 5.3 Albert Einstein in 1921, the year he was awarded the Nobel Prize in Physics (Public domain image)

university professor of physics or chemistry and not believe in the existence of atoms—you would certainly have been in a minority, but it was a position that you could simply defend on the grounds that there was no convincing experimental evidence.

Another paper proposed an explanation for the photoelectric effect: a light beam which hits a metallic surface can lead to electron emission and thereby an electric current. Einstein assumed that light was made of particles, called photons, and obtained an equation to describe the effect. Robert Millikan (1868–1953) confirmed this prediction experimentally in 1915 (which ultimately led to Einstein's Nobel Prize in Physics in 1921), while the direct experimental confirmation of the existence of photons, through what is now called the Compton effect, was provided by Arthur Compton (1892–1962) in 1923.

Finally, the last two of the five papers dealt with what is now called the special theory of relativity. (This terminology was not introduced by Einstein, and in fact he didn't particularly like it.) The word 'special' here indicates that the theory applies to the aforementioned inertial frames, where gravity is (at least apparently) absent. The extension of this theory to include the effects of gravity—which is known as the general theory of relativity—was a subsequent step, which Einstein only completed a decade later.

5.2   Special Relativity

Some historians enjoy what is known as counterfactual history, which is a discussion of how history would have unfolded if some events had turned out differently—for example, if Columbus had not found America. Although manifestly speculative, it has the merit of assessing and contextualizing the importance of a given event. Here, one could indeed say—of course, equally speculatively—that if Einstein had not existed, then someone else (most likely Poincaré, but possibly also Lorentz or others) would have arrived at special relativity in or very soon after 1905.

It should be clear from the previous section that there was a vast body of puzzling experimental evidence and theoretical analyses whose explanation required a new and perhaps radical approach, and many others were aware of and working on the problem. All that was missing was someone capable of putting all the pieces of the puzzle together and understanding how to interpret the overall pattern, and this is what Einstein provided.

Special relativity is based on a simple postulate, which Einstein called the principle of relativity. In a nutshell, this says that there are no privileged areas of physics. More specifically:

> The same laws of electromagnetism and optics will be valid for every frame of reference for which the laws of mechanics hold.

We can alternatively phrase it in terms of inertial observers by saying that all the laws of physics are the same for all inertial observers. This is therefore a generalization of Galileo's version, which only referred to the laws of mechanics.

But this seemingly harmless generalization also leads to some non-trivial consequences. The first is due to the fact that Maxwell's equations explicitly contain the speed of light, whose value is not specific to any inertial observer. Therefore, if the principle of relativity holds, this speed of light must indeed be the same for all inertial observers. In other words, faced with the choice between Newtonian mechanics and Maxwell's electromagnetism, the principle of relativity resolutely chooses the latter.

It turns out that the Newtonian law of addition of speeds, though very accurate at low speeds (as in our case of the ball in the train), is only an approximation to the relativistic result, which for two speeds v_1 and v_2 is

$$v = \frac{v_1 + v_2}{1 + \frac{v_1 v_2}{c^2}}.$$ (5.4)

The reason that this is not apparent in everyday life is that the speeds that we ordinarily deal with are many orders of magnitude smaller than the speed of light. In that case the denominator becomes, to a very good approximation, unity, and the Newtonian result approximately holds. Corrections to this behaviour only manifest themselves for speeds closer to the speed of light. For an illustration of the differences, in the simple case $v_1 = v_2$, see Fig. 5.4; the differences are only significant for speeds that are themselves a substantial fraction of the speed of light.

Notice that if in the above expression one of the speeds is that of light, say taking $v_1 = c$, then one can simplify the expression and one immediately finds

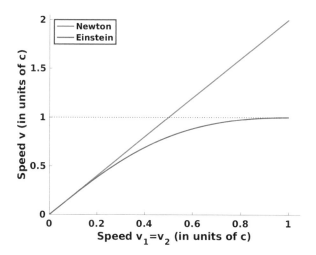

Fig. 5.4 Comparing the law of additions of speeds in Newtonian physics and in special relativity, for equal speeds ($v_1 = v_2$), as given by Eq. (5.4). Speeds are shown in units of the speed of light

that $v = c$, regardless of the value of v_2. This confirms that the speed of light is the same in every inertial frame, as required in Maxwell's electromagnetism. Notice the speed of light is the only one that is invariant—a technical term meaning that it has the same value in every frame. The speed of any material body does change from frame to frame. For Einstein—as for Galileo—there is no absolute motion.

This invariance of the speed of light also means that it is an absolute upper limit for the speed of motion of any physical object: it is impossible for anything having a non-zero mass to reach the speed of light, though it can come arbitrarily close to it, provided enough energy is put into accelerating the object to that point. Note that this does not prevent the existence of particles travelling faster than the speed of light—such particles are generically called tachyons by physicists. It does mean that tachyons can't travel at the speed of light either: they must necessarily travel faster than the speed of light, but they can come arbitrarily close to it, provided enough energy is put into decelerating them. Yes, the last part is not a typo: a tachyon with a speed very close to (but higher than) the speed of light has more energy than one with a much higher speed.

What this effectively means is that the speed of light is an impenetrable barrier. Particles with non-zero mass (or, more colloquially, massive particles) are born on either side of it and so are unavoidably confined to that side forever. The difference is that ordinary particles have real masses while tachyons have imaginary masses. On the other hand, particles with zero mass always move at the speed of light. In fact, one could coherently define the mass of a particle as its ability not to always travel at the speed of light and be able to (as it were) choose its own speed, as long as it remains on its side of the speed of light barrier.

A second important consequence of the principle of relativity is that, since it implies that the form of the laws of physics (including, of course, those of electromagnetism) is the same in all inertial frames, and all observers in constant relative motion should derive equivalent consequences from them, it follows that the laws of physics can't be used to measure an absolute velocity, and therefore the aether can't exist. If it did exist, the reference frame where it is at rest would be special: you would know that you were in it by measuring a zero relative speed (or no 'aether wind'). One consequence of this is that light requires no medium to propagate—and presumably the same should happen for the gravitational interaction.

A third important consequence is that what is known as the simultaneity of independent events becomes a relative concept. In other words, two events that are simultaneous for one observer need not be simultaneous for another.

Again, the reason we don't notice this in everyday life is that deviations from the standard (classical or Newtonian) expectation are very small when the characteristic velocities involved are small compared to that of light.

A broader consequence of the principle of relativity and the constancy of the speed of light is that the concept of an absolute space and time, as envisaged by Newton, must be wrong. Intuitively one has no significant difficulty in imagining a universal grid or cubic lattice made of equal measuring rods and fully synchronised clocks, but actually we now know that such a lattice does not exist—it's just not how the Universe works.

What does happen is that, as seen from any reference frame, a moving clock slows down and a moving rod contracts in the direction of motion (though not in the other directions), as compared to the same clock and rod at rest. These length contractions and time dilations imply that length and time intervals are actually relative quantities and depend on the state of motion of the observer that measures them. Clearly, this directly contradicts the Newtonian view of space and time (or more rigorously space and time intervals, which is what rods and clocks can measure) as absolute quantities. For a given speed v, the time dilation and length contraction factors are the same, and given by the Lorentz factor

$$\gamma = \frac{1}{\sqrt{1 - \frac{v^2}{c^2}}} . \tag{5.5}$$

Notice that this factor is always greater than unity for non-zero speeds. In other words, the relativistic length contraction formula is precisely the same as the Lorentz–Fitzgerald one, but its physical interpretation has changed significantly: it is simply a relation between the lengths of a given object as measured by observers in two different inertial frames, not a statement about an actual contraction of the length of the object as a result of its motion through the supposed aether. Figure 5.5 illustrates how the Lorentz factor depends on speed; as you might expect by now, it only deviates significantly from unity for speeds that are fairly close to the speed of light.

While the idea that space and time intervals become relative (frame-dependent) quantities seems at first counter-intuitive, upon reflection it is not much more surprising than the idea that velocities are relative. For example, assuming that you are sitting down while reading this you are at rest relative to your chair or sofa, but you are also on the surface of a rotating Earth, which is orbiting the Sun at 30 km/s. The Sun and the rest of the Solar System are orbiting the centre of our galaxy, and so on. When all these contributions are

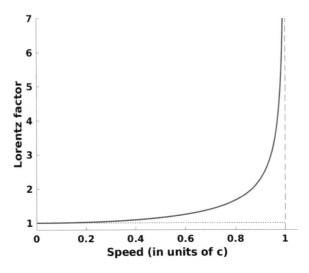

Fig. 5.5 The Lorentz factor as a function of speed, as given by Eq. (5.5). The speed is shown in units of the speed of light

put together, one finds that the total speed of the Earth with respect to the distant galaxies is more than 500 km/s. Clearly, your speed depends on the frame in which you decide to measure. What special relativity shows is that this is actually a generic feature of measurable quantities—the fact that some quantities (like the speed of light) are invariant is the exception rather than the rule.

As in the case of the law of addition of velocities, the reason why the length contraction and time dilation effects are not easily noticed is that they are typically extremely small in everyday situations—again, for small speeds the Lorentz factor is almost exactly unity. However, they can be identified in numerous experimental settings, all of which are in agreement with the predictions of special relativity. An example are cosmic rays, which keen high school students can easily detect by building their own Wilson cloud chamber.

Cosmic rays (mainly protons, but other atomic nuclei as well) constantly hit our atmosphere, coming from the Sun, from within our galaxy, or even from other galaxies. Being extremely energetic, when they collide with nuclei from the Earth's atmosphere (at an altitude of, say, 6 km) they can produce a shower of secondary particles that travel through the atmosphere and sometimes reach the ground. Some of these are muons—unstable particles whose mass is about 207 times heavier than electrons and which on average decay after about 2 ms.

So even if they were moving at the speed of light (which they are not, since their mass is not zero), one would expect them to travel only about 600 m between formation and decay. Nevertheless, they can easily be detected on the ground, even at sea level. How can they do that?

The answer is that special relativity—and more particularly the Lorentz factor—allows them to do it. Their typical speed is about 99.5% of the speed of light, and that corresponds to a Lorentz factor of about 10. From our point of view on the ground, there is a time dilation factor of 10: their clocks are running 10 times slower than ours, and they can therefore live about 20 ms (as measured by our clock) and travel for about 6 km, so they can indeed reach the ground, and be experimentally detected.

That is the view in our frame. What does the muon see? In its frame it is of course at rest, so its clock is running normally and it will decay after 2 ms. On the other hand, in its frame the Earth is moving towards it at 99.5% of the speed of light, and there is therefore a length contraction effect which as we have discussed is given by the same Lorentz factor of 10, so the Earth's surface is not initially 6 km away but only 600 m away.

The important point is that although the intervals of time and length measured in the two frames are different, the physics is the same: muons decay on average after 2 ms, and the higher we go in altitude the more of them we find (but we can still find some on the ground). Of course, we could describe this is any other frame—for example, one moving at half the speed of light with respect to the ground. That would not be a smart choice because in that frame we would have to deal both with a time dilation and with a length contraction, so the necessary algebra would be somewhat more involved. Nevertheless, the final measurable results would still be the same. In fact, special relativity is merely the set of rules of transformation between frames necessary to ensure that all observers in all inertial frames see exactly the same physics. In other words, it ensures a consistent view of the physical Universe for all such observers.

Notice that time dilation applies universally to all clocks, including biological ones: the closer to the speed of light one moves, the slower the aging process. Again the effects are very small in everyday circumstances, but when your are flying on an airplane, for example, you are aging more slowly than those who stay on the ground. A more extreme version of this is the so-called twin paradox.

Special relativity implies that space and time intervals are relative, but also that they are interlinked. In fact, one should not think of space and time as separate entities, but as a single one, which for lack of a better name one may simply call spacetime. To locate one event in spacetime one therefore needs to provide four coordinates. For example, if you're scheduling a meeting somewhere on the Earth, you should specify a latitude, longitude, altitude and time. This is why one often says—somewhat grandly—that time is the fourth dimension.

Similarly, motion should be envisaged as happening in spacetime, and not merely in space. It is then (if you will excuse the pun) relatively easy to understand the time dilation effect. If you are at rest in space you're only moving in time (at the speed of 1 day per day, say) but in order to move through space you will necessarily have to move more slowly through time.

One subtlety about spacetime is related to the concept of distance. Our ordinary concept of distance is simply a measure of how difficult it is to go from point A to B. Two such points 100 km from each other are further apart than two points only 1 km from each other. But with time intervals it is the opposite: getting from point A to B in 10 days is surely easier than doing it in just 1 day. So if I'm at point A, point B in 10 days is closer to me than point B in 1 day. So in relativity, intervals of space and intervals of time are accounted for with opposite signs when calculating spacetime distances. Special relativity makes spacetime the arena in which matter evolves, while in Newtonian theory space and time were separate entities.

Relativity also implies that the energy, mass, and speed of an object are related in the following way:

$$E^2 = p^2 c^2 + m^2 c^4 , \qquad (5.6)$$

where m is the particle mass and p is the particle momentum. For a massless particle, such as a photon, the momentum is given by

$$p = \frac{h}{\lambda} , \qquad (5.7)$$

where h is Planck's constant and λ is the wavelength, while for a massive particle we have

$$p = m\gamma v , \qquad (5.8)$$

where v is the speed and γ is again the Lorentz factor. Thus the energy of a massive object increases with its speed, while that of a massless one increases with its frequency (which is inversely proportional to its wavelength). Also, for the case of speeds much smaller than the speed of light, Eq. (5.6) can be approximately written

$$E = mc^2 + \frac{p^2}{2m} + \cdots, \tag{5.9}$$

with the first and second terms being the rest energy and the kinetic energy.

Note that, since in Eq. (5.8) the particle mass is multiplied by the Lorentz factor, one could interpret the product of the two quantities by saying that the particle mass depends on speed—and also calling m the rest mass (meaning the mass of the particle when it is not moving). This may be a useful mnemonic because it provides a simple way of understanding why a massive particle can't reach the speed of light: as its energy increases so does its mass, and therefore it becomes harder and harder to accelerate—so much so that reaching the speed of light would require an infinite amount of energy. However, this is still based on a Newtonian interpretation. Mass is a relativistic invariant, while energy does change in different inertial frames.

But special relativity deals only with inertial reference frames, while Nature does allow for non-inertial (accelerating) observers. To deal with them, general relativity is needed. With this generalization spacetime itself will become dynamical. How this happens is what we will now discuss.

5.3 The Equivalence Principle and General Relativity

In the previous section we mentioned that Einstein introduced special relativity in the context of a plethora of experimental results (many of them contradicting others) and various attempts by others to provide the presumed missing theoretical framework. So, had Einstein not existed someone else would presumably have arrived at special relativity in or nor long after 1905. It is therefore interesting that general relativity emerged in almost exactly the opposite way.

In this case there was no substantive experimental motivation—no puzzling experimental result that somehow conflicted with current knowledge or called for a deeper examination of underlying theoretical concepts. Moreover, Einstein developed it almost entirely alone, at least when it comes to the physics aspects. On the mathematical side, Einstein did have the crucial help of his colleague and friend Marcel Grossmann (1878–1936), from whom he learned the necessary mathematical techniques—today generically known as tensor calculus—as well as the importance of non-Euclidean geometries.

Einstein's starting point was a set of superficially simple questions such as what makes inertial frames special. One peculiarity is that in such frames, for instance, in an elevator whose cable breaks (or, perhaps less dramatically, when we dive from a platform into a swimming pool), we don't feel gravity. This may seem an innocuous observation, but it turns out to be very significant—and part of Einstein's genius was in understanding precisely why this is so.

The first breakthrough came in 1907, only 2 years after the development of the special theory of relativity. Einstein had of course realised in 1905 that Newton's gravitation was incompatible with special relativity, and therefore looked for a way to modify it. The incompatibility resides in something that has already been mentioned: Newton's laws have no explicit propagation speed. So if something happened to the Sun, for example, the Earth would instantaneously feel the corresponding gravitational effects. But a physical effect travelling at infinite speed—or indeed at any speed faster than that of light—is incompatible with special relativity. So Newton's laws can't be correct, although presumably they may be a reasonable approximation in some special set of circumstances.

It follows from Galilean inertia that no experiment in a closed box (think of a train, for example) is able to absolutely determine the state of uniform motion of the box. This is sometimes anachronistically called the Galiean equivalence principle. Einstein generalised this in two different but inter-related ways:

No local experiment in a small closed freely falling box can determine whether the box is falling in a gravitational field or simply floating in space, far away from any gravitational forces.

No local experiment in a small closed box can determine whether one is at rest in a uniform gravitational field of a given magnitude (such as on the surface of a planet) or in a rocket being uniformly accelerated through empty space (with an acceleration of the same magnitude).

If we drop a stone and it falls, we can say that this is due to its being in a gravitational field, but we can equally say that the ground of the Earth (or the floor of a spaceship) is accelerating towards it. Having grown used to gravity on the Earth it may be hard to believe that the second explanation is equivalent to the first, but in Einstein's Universe this is true. The reason for this is that the acceleration of any body in the presence of a given gravitational field is the same. This is called the principle of equivalence—or more specifically the Einstein equivalence principle (EEP)—and it is the key concept of general relativity.

One of the immediate consequences of this principle is that the inertial and gravitational masses of any object are the same. Recall that in Newtonian physics there is an analogous Newtonian equivalence principle (though naturally, the term is equally anachronistic, never having been used by Newton himself), which states that the two masses must be proportional to each other, but the proportionality factor is undetermined and can be freely chosen. In Einstein gravity there is no choice: if the EEP holds, the two must be identical.

A more subtle but equally important consequence is that, given any gravitational force, one can always choose a reference frame in which an observer will not feel any gravitational effects in its neighbourhood—such a frame is said to be freely falling. Although this frame can continuously change with time, there is always one such frame. What this means is that gravity is actually an apparent or fictitious force: one can always locally switch it off by placing oneself in the appropriate (freely falling) frame. Note that it is not possible to do the same with electromagnetic forces. As we will see in more detail in a moment, this apparent force is due to the local geometry (or curvature) of spacetime, which in turn is due to its matter content.

For an observer in free fall, there is apparently no gravitational field, at least in its immediate neighborhood. Dropping any objects, they will remain at rest relative to the observer, or will move with uniform motion. This is what happens, for example, with astronauts inside the ISS, as we have already discussed in the previous chapter. Such observers can therefore legitimately claim to be at rest.

The universality of free fall (that is, the fact that all bodies in the same gravitational field will fall with the same acceleration) becomes a simple corollary of the EEP. Indeed, one could make the reverse argument: the empirical observation that all objects fall in the same way in a given gravitational field supports the view that gravity is a purely external effect, due to the geometry of spacetime itself and not to the specific properties of each individual object.

Note that so far we have only discussed the Einstein equivalence principle (which Einstein obtained in 1907), and not general relativity (which he only obtained in 1915). That's because the two are conceptually different. The EEP states that gravity is only geometry (or that spacetime is curved by the presence of matter) but it does not state how much curvature is generated by, say, one kilogram of matter. Theories satisfying the EEP are generically called metric theories of gravity, and there are in principle an infinite number of them. Each of these will predict a different number for the amount of curvature generated by one kilogram of matter.

General relativity is one of these theories, and makes its own specific prediction for this number. Thus Einstein first reached the key physical concept, in 1907, but at that point he was still lacking the necessary mathematical background to fully develop his theory and carry out a detailed exploration of its consequences. This is where Marcel Grossmann's help came in, as mentioned previously.

In Newtonian gravity, space and time are a fixed arena which contains matter. Each of these pieces of matter produces a gravitational field that attracts all matter, and indeed these effects are propagated and felt instantaneously. In Einstein's gravity, matter and spacetime are no longer disconnected, but they interact:

> Space tells matter how to move.
>
> Matter tells space how to curve.
>
> John Archibald Wheeler (1911–2008)

Gravity is therefore a property of spacetime, whose properties it also affects. This relation between the two is mathematically encoded in what came to be known as Einstein's field equations. In particular, these equations enable us to quantitatively determine how much spacetime curvature is produced by one kilogram of matter. A larger mass naturally leads to greater curvature.

The motion of each object is determined by the local curvature of spacetime in the region where the object happens to finds itself. If this curvature happens to be small, the behaviour predicted by Einstein's theory turns out to be the same as the Newtonian one, but otherwise the predicted behaviours of the two theories are different—and sometimes dramatically so. These differences can and have been experimentally identified, as we will discuss at the end of this chapter.

As a slight epistemological digression, it is worthy of note that, although general relativity and Newtonian gravity are mathematically related, in the sense that one can start with the dynamical equations for the former and, taking an appropriate mathematical limit, recover the latter, the two are not physically related—in the sense that the underlying physical interpretation associated with the two is not only different but mutually incompatible. In that sense general relativity is not an innocuous extension of Newtonian gravity: it implies a thorough change in our view of the Universe and its workings. More-over, the two views can be experimentally and observationally distinguished, because they do lead to different predictions in many circumstances, as we will discuss in the next section.

This is one of the many examples showing that mathematics and physics are conceptually different things. The fact that one mathematical formula may apply to two different physical contexts does not in any way imply that the physical interpretation of that formula is the same in the two contexts; there is no reason why it should be so, and in some situations the interpretations are demonstrably different. We will discuss another good illustration of this point later in the book.

Gravity also affects time—or, more to the point, how clocks measure time. We have already discussed this in the context of special relativity, specifically the time dilation effect, but general relativity leads to an additional effect: the greater the intensity of a gravitational field, the more slowly time flows. Again, this effect has been confirmed very accurately. For example, we age more slowly in a gravitational field than someone in a gravity-free environment. In fact, for the same reason, when you are standing up, the cells in your brain are aging slightly faster than those in your feet.

The development of general relativity is one of the most important achieve-ments of the human mind—all the more so because it was largely due to a single person. The view of the Universe provided by general relativity (and, more generically, the EEP) is radically different from the previous Newtonian one. One of the difficulties that many people have when they first encounter relativity is that a lot of the Newtonian view makes sense in the context of our everyday intuition—but remember that many people would have said the same about Aristotelian physics a few centuries ago.

With Einstein's gravity spacetime is no longer the arena in which the cosmic drama unfolds, as was the case in special relativity. Instead, it is now part of the cast: spacetime properties determine the way in which objects move, but these properties are themselves determined by the objects that it contains. In other words, spacetime itself is dynamical, whereas it was static in Newtonian gravity. This historical evolution of our concepts of space and time, which were initially unrelated and absolute, then became a single entity, and finally acquired their own dynamics due to the contents of the Universe, is a profound conceptual transformation.

During the last century general relativity and the EEP have passed a vast set of experimental and observational tests, a small number of which will be discussed in the next section. Nevertheless, many scientists believe that they both have a limited range of applicability—although that range is certainly far greater than the one for Newtonian gravity—and therefore that they are doomed to be replaced eventually by an even deeper and more fundamental theory. We will discuss some possibilities along those lines later in the book.

5.4   Gravity Is Geometry

One of the most counter-intuitive concepts to grasp when first encountering general relativity is the idea that gravity is an apparent force which is the result of the geometry (specifically, the local curvature) of spacetime. There are several classical and relatively simple ways to illustrate the idea, which are worth knowing and keeping in mind.

Let us consider a disk whose radius R and perimeter P we will measure with a ruler. Naturally, we will find that the two are related by $P = 2\pi R$. Now let's set the disk in motion, specifically rotating about its perpendicular axis with some constant angular speed. Note that for the perimeter, each section of the edge of the disk will be measured by our ruler along its direction of motion, and will therefore be subject to length contraction in accordance with special relativity. The perimeter of a rotating disk will therefore be smaller than that of the static one. On the other hand, when we measure the radius the measurement is always done perpendicularly to the direction of motion, and therefore to the disk's velocity; in this case there is no length contraction and therefore the measured radius is the same as in the case of the static disk. So for a rotating disk $P \neq 2\pi R$. How can this be?

The reason is that the standard relation between radius and perimeter only applies to standard Euclidean geometry. You can easily convince yourself of this by drawing circles on the surface of a sphere and measuring all distances—that is, both for the radius and the perimeter—along the sphere's surface. In that case you will find that the perimeter is smaller than 2π times the radius. We can thus say that the length contraction effect makes a rotating disk have a geometry that is akin to that of a sphere. Note that this is purely an effect of special relativity. We have not invoked general relativity.

Let us additionally assume that the Einstein equivalence principle holds, and imagine that we are inside a small closed box on the periphery of the rotating disk. Clearly, we know that we are in an accelerated frame, since we feel a force: remember that a rotating object is always changing its velocity (in direction, though not in magnitude) and it is therefore accelerating. If the EEP is valid, no experiment in our laboratory box can distinguish between a gravitational force and an accelerated system. Thus we can legitimately say that our acceleration is due to a gravitational field, which in turn is due to the non-Euclidean geometry of the rotating disk.

For a different but analogous example, imagine that you and a friend start out some distance apart on Earth's equator and travel directly north, in initially parallel trajectories. As you do so the distance between the two of you will decrease (and eventually you will meet up at the North Pole). You can describe your trajectories by saying that you are moving on a locally straight line (unaffected by any forces) on the surface of a sphere, but if you did not know that you were on the surface of a sphere you could equivalently say that you were moving on a flat surface but that there was an attractive force between the two of you that caused your distance to decrease as you moved.

Finally, consider two apples falling straight down, from a great altitude, towards the centre of the Earth. An observer accompanying their fall will notice that the distance separating them decreases as they fall. Again we can say that a gravitational force makes the apples approach each other, but we can also say that it is the spacetime curvature, in this case caused by the Earth, that does this. If the EEP holds true, the two explanations are identical and indistinguishable—no experiment can possibly allow you to separate one from the other.

In our everyday life—say on Earth, or even throughout the whole Solar System—the gravitational fields are comparatively weak. In this case general relativity can be thought of as providing some relatively small corrections to the behaviour predicted by standard Newtonian theory. These are usually called the post-Newtonian corrections, which are of 3 types:

Curvature effects: the space through which planets, asteroids, or comets move is curved, so the distances and angles will also be slightly different from the ones that would hold in flat spacetime.

Velocity effects: since all energy (and not just mass) gravitates, the effective increase in the inertial mass of moving bodies leads to a modification of the gravitational force.

Nonlinear effects: in Newtonian theory the gravitational force can be expressed as being proportional to another quantity, called the potential, but in general relativity there is an additional term proportional to the square of this potential (with the opposite sign and a small coefficient), meaning that the overall force is slightly smaller.

Despite the smallness of these effects in everyday circumstances, several observations supporting general relativity, both in the Solar System and elsewhere, have been accumulating over the past 100 years. It is worth keeping in mind that, since the EEP and general relativity are conceptually different (as was emphasised in the previous section) some observations test the former (that is, whether gravity is purely geometry) while others test the latter (that is, specifically how much curvature is generated by one kilogram of matter). The difference is important: if some test happens to show that the EEP is violated that will imply that general relativity is incorrect, but if some test shows that general relativity is incorrect that does not imply that the EEP is violated—it is possible that another metric theory of gravity is the correct one.

5.5   Consequences and Tests

Let's start close to home, with the satellites of the GPS system. We have mentioned in the previous chapter that they could not work without relativity, and now it is easy to understand why. There are two relativistic effects on the clocks on board the satellites, as compared to otherwise identical clocks on the ground. General relativity implies that clocks on the satellites, being in a weaker gravitational field than that on the Earth's surface, will run faster. The difference is about 45,900 ns per day. On the other hand, special relativity predicts that the satellite clocks, which are orbiting the Earth, should run slower than those at rest, the difference in this case being about 7200 ns per day. The overall effect is therefore about 38,700 ns per day.

Now, you will remember that we mentioned in the previous chapter that the 10 m precision of the civilian version of the GPS required a clock precision of 30 ns. Thus the relativistic corrections are huge on this scale: a Newtonian version of the GPS would only reach a precision of 13 km after 1 day (and 26 km after 2 days, and so forth). In fact the situation is even more complex because the Earth's gravitational field is not uniform: it is different over Mount Everest and over the Mariana Trench, for example. So these relativistic corrections continuously change, and are different for each satellite. In fact, nowadays, one can make the reverse argument: assuming that relativity is correct one can predict the behaviour of the clocks, and any discrepancies between the predictions and their actual behavior should be due to variations of the Earth's effective gravitational field felt by the satellites, which can therefore be reconstructed.

The first test of general relativity was in fact provided by Einstein himself, and we have already discussed it in the previous chapter. This was his calculation of the relativistic correction to the amount of precession of Mercury's perihelion, as compared to the Newtonian result, which yielded precisely the missing value. Using the modern values the prediction of general relativity is 42.98 arcseconds per century, while the measured value is 42.98 ± 0.04 arcseconds per century.

One of the more interesting consequences of general relativity and the EEP is that, since light has energy, it also gravitates—in other words, photons are attracted by gravity, and their trajectories are bent by gravitational forces. If a photon is emitted near a massive body such as a planet, its trajectory will be slightly curved towards the planet. One might think that this implies that such trajectories are no longer straight lines, and this would be true if the space in which the photons are moving were the usual, Euclidean one. However, the planet's mass curves the space around it; photons are still moving in straight lines, but these are defined in curved spacetime—mathematicians call these trajectories geodesics. Notice that in this description we no longer make use of a gravitational force, so gravity is indeed a superfluous concept.

As you can by now guess, this effect is also tiny in everyday circumstances, but its observation in the total solar eclipse of 29 May 1919 (see Fig. 5.6) provided the second piece of supporting evidence for of general relativity, and the one that made Einstein an instant celebrity. For the case of the Sun, the deflection predicted by general relativity, for the case of a photon trajectory grazing the Sun, is 1.75 arcseconds.

This prediction comes from two equal-sized parts. One half comes from the EEP, and therefore it also exists in Newtonian gravity: one can think of it as the effect of the weight of light, and can be calculated using Newtonian gravity for a particle that happens to be moving at the speed of light. In fact, such a calculation was done by Henry Cavendish (1731–1810) around 1784 and by Johann von Soldner (1776–1833) in 1803. But the other half is new, and it is a geometric contribution, from the fact that mass curves spacetime, and it is therefore specific to general relativity. Other metric theories of gravity will make different predictions for this second contribution; thus to some extent what distinguishes general relativity from other metric theories is that it is the one for which the contributions of the two effects are numerically identical.

The person who recognised the significance of the 29 May 1919 eclipse (which had the Sun in the direction of the Hyades cluster, a compact group of conveniently bright stars) was Frank Dyson (1868–1939), who was Astronomer Royal at the time, while Arthur Eddington (1882–1944), being already familiar with general relativity, saw the opportunity to test it. Eddington and Edwin Cottingham (1869–1949) carried out the measurements at Principe island (then part of Portugal), while Charles Davidson (1875–1970) and Andrew Crommelin (1865–1939) who were Dyson's assistants at the Greenwich Observatory at the time did the measurements at Sobral (in Brazil).

Note that apart from Einstein's and Newton's predictions there was in principle a third possibility: if light did not gravitate there should be no deflection at all. Thus the 1919 eclipse expedition results were doubly important: first

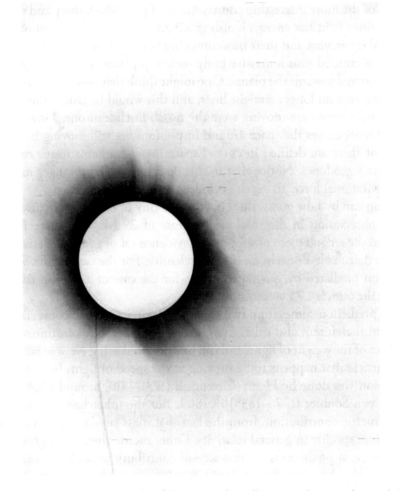

Fig. 5.6 One of the photographs of the 1919 solar eclipse experiment, taken at Sobral, presented in his 1920 paper by F.W. Dyson, A.S. Eddington, and C. Davidson, Phil. Trans. Royal Soc. Lond. A **220** (1920) 291 (Public domain image)

they demonstrated that light does gravitate, and second they showed that the Newtonian prediction was incorrect (while being compatible with Einstein's prediction). The contemporary media understandably emphasised the latter (see Fig. 5.7), but the former was equally important.

This effect of light bending can also be observed on a grander scale across the Universe, in the so-called gravitational lenses. This is something that Einstein himself had predicted in 1937. The first was discovered in 1979, and nowadays they are quite common. In addition to being a source of beautiful pictures (see

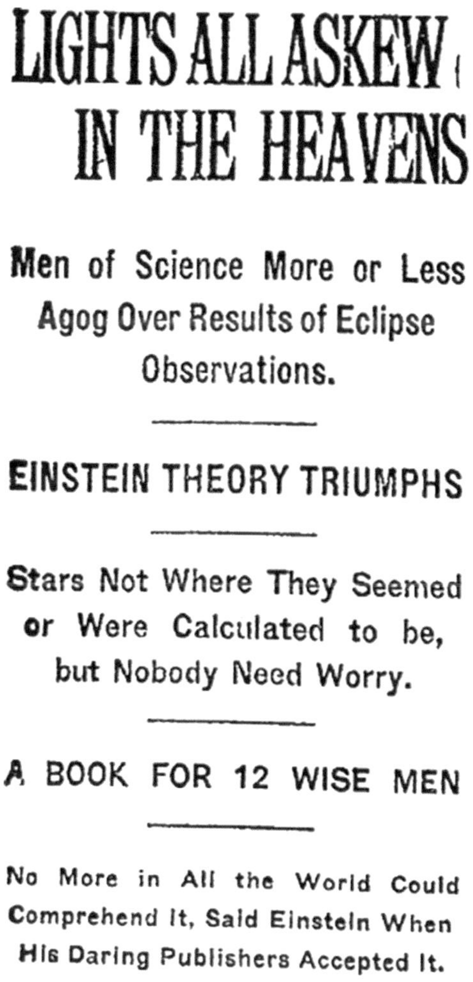

LIGHTS ALL ASKEW
IN THE HEAVENS

Men of Science More or Less Agog Over Results of Eclipse Observations.

EINSTEIN THEORY TRIUMPHS

Stars Not Where They Seemed or Were Calculated to be, but Nobody Need Worry.

A BOOK FOR 12 WISE MEN

No More in All the World Could Comprehend It, Said Einstein When His Daring Publishers Accepted It.

Fig. 5.7 Part of page 17 of the New York Times of 10 November 1919, reporting the results of the 1919 eclipse expedition (Public domain image)

Fig. 5.8), they can also be used for remarkably interesting science, including measuring the amount of dark matter in the Universe and its rate of expansion.

Another decisive prediction of general relativity is the existence of gravitational waves, which can be thought of as waves of spatial curvature propagating at the speed of light. Again these are tiny, and therefore extremely difficult to detect directly. Strong indirect evidence for their existence has been known for several decades due to a so-called binary pulsar, which is slowly losing energy to gravitational waves and thereby decreasing its orbital period. Radio

Fig. 5.8 The 'Smiley' image of galaxy cluster SDSS J1038+4849, with arcs (known as an Einstein ring) caused by strong gravitational lensing. Credit: NASA/ESA

telescopes can measure this orbital period (and its change) very accurately, and this is in excellent agreement with the prediction of general relativity. More recently, these gravitational waves have been detected directly by the LIGO and Virgo projects. Interestingly, these detectors are essentially sophisticated modern versions of the Michelson interferometers used in the Michelson–Morley experiment.

We have been discussing tests of general relativity, but let's finish by discussing one test of the Einstein equivalence principle (we will discuss others later in the book). This is called the gravitational redshift, which Einstein already predicted in 1907—the year in which he arrived at the EEP itself.

Consider a tower somewhere on the Earth, containing a device that can emit a beam of light towards the ground, where it will be picked up by a receiver. An observer happens to be on a freely falling elevator at the same height above the ground as the emitter and at the precise moment then the beam of light itself is emitted, so the two will fall in the same way. Being in free fall, the observer can describe what will happen to the beam of light using only special relativity.

For the observer the frequency of the beam of light is unchanged throughout the fall, but the ground (and therefore the receiver) are moving towards the elevator and the beam of light. Therefore the receiver will detect a blue shift (that is, a higher frequency) due to the Doppler effect, determined by the speed of the elevator as seen from the ground (or the receiver as seen from the elevator) at the moment of detection. If we had an opposite set up, with the emitter at the bottom and a receiver at the top of the tower, we would analogously predict a redshift, since in that case the receiver would be moving away from the observer in the elevator.

Now, if instead of having an emitter which sends a single beam of light we have one that sends them periodically—in other words, if we have a clock— we infer that an observer on the ground would see that the clock sitting at the top of the tower was moving faster than an otherwise identical clock on the ground. (In fact this is precisely the effect on the GPS clocks that was mentioned earlier.) Thus in this case we would have a blueshifted clock rate, while a clock on the ground would have a redshifted rate. This is the origin of the term gravitational redshift.

Why do the clock rates change? Is it something to do with the emitter and receiver, or does something happen to the light along the way? This question is analogous to asking why do lengths contract and clocks slow down in special relativity. As you might remember, Lorentz tried to obtain a physical mechanism underlying its transformations, invoking some molecular force that would cause the contractions. In a sense special relativity bypasses the question, or rather it answers that it doesn't matter. All one knows is the lengths and time intervals that are measured in each frame, and how they relate to one another depending on the relative velocity of the frames.

Here, the answer is analogous. What general relativity (and, remember, all other metric theories of gravity) can predict is how the two clocks will differ, depending on how the gravitational field varies between the two altitudes. One could try to describe what happens in terms of changes to the emitter and receiver or in terms of changes in the trajectory (do try it!), but the two descriptions will be the same since there is no way to distinguish between them experimentally.

It goes without saying that, as in the case of special relativity, this affects all clocks—whether they are mechanical, electromagnetic, atomic, or biological. The fact that this is a generic prediction of the EEP can be seen from the fact that general relativity was not used at all in the above paragraphs. It is a consequence of the EEP (we did have to assume that the elevator and the beam of light were falling in the same way), and therefore it simply tests whether gravity is indeed purely geometrical. The quantitative amount of gravitational redshift predicted by general relativity and all other metric theories of gravity is exactly the same.

6

The Physics of Stars and Galaxies

In this chapter we will be considering in detail the astrophysical building blocks of the Universe, stars. We first overview their main physical properties and life cycle, including their three possible end states: as white dwarfs, neutron stars, or black holes. We also briefly highlight their energy production mechanisms, including their role in the formation of the elements in the periodic table. Finally, we take a brief look at how we came to discover that we are part of a galaxy (the Milky Way) that is one of countless many throughout the Universe, and moreover that the Universe is not static but is expanding.

There are countless suns and countless earths all rotating round their suns in exactly the same way as the seven planets of our system. We see only the suns because they are the largest bodies and are luminous, but their planets remain invisible to us because they are smaller and non-luminous. The countless worlds in the Universe are no worse and no less inhabited than our Earth. For it is utterly unreasonable to suppose that those teeming worlds which are as magnificent as our own, perhaps more so, and which enjoy the fructifying rays of a sun just as we do, should be uninhabited and should not bear similar or even more perfect inhabitants than our Earth. The unnumbered worlds in the Universe are all similar in form and rank and subject to the same forces and the same laws.

Impart to us the knowledge of the universality of terrestrial laws throughout all worlds and of the similarity of all substances in the cosmos! Destroy the theories that the Earth is the centre of the Universe! Crush the supernatural powers said to animate the world, along with the so-called crystalline spheres! Open the door through which we can look out into the limitless, unified firmament composed of similar elements, and show us that the other worlds float in an

(continued)

© Springer Nature Switzerland AG 2020
C. Martins, *The Universe Today*, Astronomers' Universe,
https://doi.org/10.1007/978-3-030-49632-6_6

ethereal ocean like our own! Make it plain to us that the motions of all the worlds proceed from inner forces and teach us in the light of such attitudes to go forward with surer tread in the investigation and discovery of Nature! Take comfort, the time will come when all men will see as I do.

Giordano Bruno (1548–1600)

6.1   The Life of a Star

From a physics perspective stars are interesting because all the four fundamental forces of Nature are relevant for their properties and evolution, although gravity is the one that plays the deciding role and will ultimately determine what happens in the final stage of the star's life cycle.

To a first approximation, the stars we see in the night sky are characterised by two basic physical properties, although these properties are perhaps not those that one might think of as the most natural ones. The first of these is the intrinsic luminosity: the total amount of electromagnetic energy emitted by the star per unit of time. (Note that this includes all electromagnetic radiation, that is in all wavelengths, and not just those in the part of it that our eyes can see.) Terms such as brightness or magnitude are ways to quantify how bright a star appears to us, but that will depend on its distance and also (if we are looking at it with our own eyes) on how much energy it emits in the visible part of the electromagnetic spectrum.

Note that magnitude is conventionally defined such that a smaller (and possibly negative) number means a brighter object. This convention comes from the Hellenistic period: Hipparchus and Ptolemy classified the stars visible to the naked eye in six different classes, with the first magnitude ones being the brightest and the sixth magnitude ones being the faintest. The modern definition builds upon this one, using a logarithmic scale such that a difference of one magnitude corresponds to a difference in brightness of a factor of the fifth root of 100, which is approximately 2.512. In other words, a first magnitude star is one hundred times brighter than a sixth magnitude star. The apparent magnitudes of some astronomical objects are listed in Table 6.1.

The second basic property is their colour, defined in the physics sense as the wavelength at which the star emits most of its energy. In other words, the distribution of the energy emitted as a function of wavelength follows a particular curve (known to physicists as a black-body curve) and the wavelength corresponding to the peak of that distribution provides the colour.

Table 6.1 Approximate apparent magnitudes of some astronomical objects

Apparent magnitude	Example(s)
−27	Sun
−13	Full Moon
−12	Betelgeuse, when it becomes a supernova (estimated)
−9	Brightest artificial satellites (Iridium flares)
−6	Maximum brightness of the ISS
−5	Venus (maximum)
−4	Faintest objects visible with the naked eye around noon
−3	Jupiter and Mars (maximum)
−2	Mercury (maximum)
−1	Sirius (the brightest star)
0	Saturn (maximum), Vega, Arcturus
1	Antares
2	Polaris (the pole star)
3	Supernova SN 1987A
3	Andromeda galaxy
5	Uranus (maximum), Vesta (brightest asteroid, maximum)
6	Typical naked eye limit
7	Ceres (dwarf planet, maximum)
8	Neptune (maximum)
10	Typical limit for standard binoculars
11	Proxima Centauri (nearest star after the Sun)
14	Pluto (dwarf planet, maximum)
27	Visible wavelength limit of ESO's VLT
32	Visible wavelength limit of the HST

The numbers have been rounded off to the nearest integer

Since the way a star emits energy is close to a black body, the colours of stars seen by our eyes are approximately those of a black body with the corresponding temperature. Therefore the coolest stars look deep red (although the peak of their energy emission can be well into the infrared part of the spectrum), and hotter stars are successively reddish orange, yellow, white, and light blue. Anyone who has reasonably good eyesight and experience of stargazing should be able to remember seeing examples of stars of each of these colors. Remarkably, we do not see any green stars, although there are a few stars that look greenish, at least to some people. Often this is an optical illusion: a red object can make nearby bluish objects look green.

The reason why we can't see truly green stars is that a star whose peak of emission is in the green part of the electromagnetic spectrum will necessarily also emit significant amounts of energy in the red and in the blue parts of the spectrum, so to our eyes (and brain) it will look white. The reason that the hottest stars look bluish (rather, as you might have expected, violet) is that for those the peak of emission is well into the ultraviolet, and the amount of energy emitted in the visible part of the spectrum is significant throughout this range, though skewed towards the blue part.

The intrinsic luminosity and colour are the two main properties of a star in the sense that other important properties can (at least to a very good approximation) be obtained from them. In particular, the mass can be obtained from the luminosity, and the temperature can be obtained from the colour. On the other hand, the radius can be obtained if both the luminosity and the colour are known.

The two-dimensional plot showing the stars' luminosities versus colours is known as the Hertzsprung–Russell diagram (see Fig. 6.1), proposed independently around 1910 by Ejnar Hertzsprung (1873–1967) and Henry Norris Russell (1877–1957). Its relevance stems from the fact that, plotting large numbers of stars therein, one notices that they are not uniformly spread in the diagram but occupy specific regions of it. In particular, throughout most of their lives, stars occupy a diagonal strip in the diagram, known as the Main Sequence.

Any cloud of gas left to the sole influence of gravity would gradually collapse due to the mutual attraction of its particles, and this would also apply to a star, which to a first approximation is a very large cloud of hydrogen. Preventing this collapse requires a further physical mechanism to balance the effect of gravity. In the case of stars this is an outwards pressure, the energy source being nuclear fusion reactions taking place in the inner part of the star, known as the star's core. As we will soon see, this mechanism can work only for a limited amount of time—how long this happens to be will mainly depend on the star's mass (or equivalently, its luminosity).

In order to study the formation and evolution of stars, Newtonian physics provides fairly satisfactory answers, as long as a little bit of quantum mechanics is added in. However, to understand their final stages one must go beyond Newtonian physics. For white dwarfs one needs both quantum mechanics and special relativity, while general relativity is not essential. On the other hand, for neutron stars and black holes the situation is different: they can only be understood in the context of general relativity.

The life cycle of a star starts with a cloud of hydrogen, also containing some small amounts of dust and other debris, possibly coming from the leftovers of an earlier generation star. Some density perturbation, whether its origin is primordial (from the early Universe, as we will discuss in the next chapter) or local (from a nearby supernova explosion) will affect the cloud and trigger its collapse. A region that is slightly denser than the average density in its neighbourhood will have a stronger than average gravitational field and therefore it will attract some neighbouring matter, thus becoming

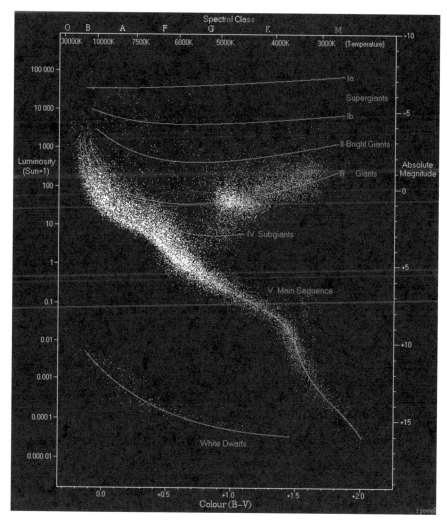

Fig. 6.1 A Hertzsprung–Russell diagram, containing 22,000 stars from the Hipparcos Catalogue together with 1000 low-luminosity stars from the Gliese Catalogue of Nearby Stars. Credit: Richard Powell

denser, attracting even more matter, and so on. In other words, since gravity is always attractive, once this collapse starts it is unstoppable in the absence of competing forces, although it will take a long time, since gravity is a comparatively weak interaction.

As the cloud is contracting, it will also be rotating progressively faster, much in the same way as an ice-skater can turn faster by pulling her arms and hands closer to her body. It is also becoming more and more compact, with its central density increasing with time. After a sufficient time—say, around one billion years—we will have a reasonably defined primitive planetary system, composed of a proto-star (already quite big, but still too cold to start nuclear reactions) and quite likely also some proto-planets which were circling the proto-star sufficiently fast to avoid falling into it (think of an artificial satellite orbiting the Earth), but instead became centres of gravitational attraction in their own right. The proto-star and any proto-planets will be embedded in an envelope of dust and gas, some of which is still gradually falling onto the proto-star. As the proto-star and the proto-planets become more compact, their internal temperature also increases. At this stage the proto-stars can already be hot enough to emit a significant amount of infrared radiation. Unlike visible light, infrared radiation can get through the dust envelope, and therefore this stage of stellar evolution can be observed and studied.

Assuming that the proto-star mass is high enough—above about eight percent of the mass of the Sun, or 80 times the mass of Jupiter—its central temperature will eventually reach about ten million degrees, at which point nuclear reactions can occur efficiently in the inner part of the star, known as its core. The energy thus released not only increases the star's temperature but also increases the pressure, and temporarily stops the process of gravitational collapse. The effects of this outwards pressure continue outside the star itself (in the case of the Sun, this is known as the solar wind), causing most of the leftover debris from the original cloud to be expelled from the system; the planets themselves also contribute to clean up the newly formed stellar system.

Incidentally, our galaxy's current birth rate is about seven stars per year, but younger galaxies with sizes comparable to ours may produce several thousand stars per year. The upper limit for stellar masses is not at all well known: it may be as low as about 120 solar masses, but there are contested claims of discoveries of stars with higher masses. This 120 solar mass limit is referred to by astronomers as the accretion limit, the physical argument for it being that beyond this mass the star is so hot that it will evaporate away any new material as quickly as it falls in. One reason for this uncertainty on the upper limit is that the distribution of stellar masses is very heavily skewed: low-mass stars are very common, while high-mass stars are rare.

As a small digression, proto-stars below the limit of 80 Jupiter masses are known as brown dwarfs. The lower end of their masses is not well known but is thought to be around 13 Jupiter masses. In other words they span the range between the most massive planets and the lightest stars. Brown dwarfs are unable to fuse hydrogen into helium, but it has been suggested that they obtain some of their energy by fusing deuterium (a heavier isotope of hydrogen). From their name you might think that if you saw one it would look brown,

Fig. 6.2 Comparison of the size of a typical brown dwarf with those of the Sun, Jupiter, the Earth, and a low-mass star. Credit: NASA/JPL-Caltech/UCB

but this would not necessarily be the case. Their colour is expected to depend on their mass, and in fact many of them would look reddish or magenta to the human eye (see Fig. 6.2).

Getting back to ordinary stars, upon reaching the Main Sequence the star finds itself in a stable plateau in its evolution, though still a finite one. Depending on the star's mass, it may stay in this phase for anything between only a million and more than ten trillion years. Surprisingly, the more massive a star is, the shorter will be its life. A heavier star has a stronger gravitational field to be balanced, and although it has more nuclear fuel (hydrogen) available to burn, it will burn it so much more rapidly that it will run out of it far sooner than a smaller mass star. In a star like the Sun this stage lasts about ten billion years, and the Sun is currently about half way through this stage. Some representative properties of stars of different masses are summarised in Table 6.2.

Eventually, the supply of hydrogen in the star's core will become depleted; the nuclear reactions will no longer occur efficiently and they will soon stop. An immediate consequence of this is that the pressure drops and the gravitational collapse of the star, which had been interrupted by the onset of the Main Sequence, will now continue. As the core contracts and becomes denser its temperature also keeps increasing until, at a temperature of about

Table 6.2 Basic properties of stars of different masses and luminosities, given in units of those of the Sun

Mass	Luminosity	Surface temperature	Main Sequence time
25	80,000	35,000	0.003
15	10,000	30,000	0.015
3	60	11,000	0.5
1.5	5	7000	3
1	1	6000	10
0.75	0.5	5000	15
0.5	0.03	4000	200
0.2	0.006	3100	1500
0.1	0.0009	2600	6000
0.08	0.0003	2400	11,000

The surface temperatures are in degrees kelvin and the times in the Main Sequence in billions of years (with one billion meaning one thousand million)

100 million degrees, the burning of helium into carbon (and also a bit of oxygen) becomes efficient. This new source of energy further raises the core temperature, while the radiation pressure will push out the external layers of the star.

The result is a star which, as seen from the outside, will be many times larger than its previous size. Its inner structure has also changed: the star now has a smaller and much hotter core (where nuclear reactions are taking place), but this is surrounded by larger outer layers—containing perhaps as much as twenty percent of the star's original mass—which will be cooler than in the previous Main Sequence phase and will therefore shine red. This is the origin of the common name for these stars: they are known as red giants. This red giant phase is fairly short, both because the energy production rate is now faster and because the amount of available helium is small. Typically, this phase lasts, for each star, about one percent of the corresponding hydrogen-burning phase. Our Sun will go through this red giant phase in about 4.5 billion years. When this happens it will grow so large that will span the orbits of Mercury and Venus, and possibly even reach that of the Earth.

But as in the case of hydrogen, this helium burning phase can only be kept up for a limited amount of time. Eventually, the supply of helium in the star's core will run out, and the previous story will repeat itself: the gravitational attraction has no competition, and therefore the star collapses further. For sufficiently massive stars other nuclear reactions can subsequently become viable—for example, carbon and oxygen can burn at a temperature of at least 3000 million degrees—so the energy output will increase for another limited period until the various nuclei are depleted, and the contraction takes

Table 6.3 The typical duration of the successive phases of burning of several elements, and the corresponding typical temperatures (in degrees kelvin) for a star with 15 solar masses

Fuel	Main products	Time	Temperature
H	He	Ten million years	Five million
He	C	Few million years	100 million
C	O, Ne, Mg, He	1000 years	600 million
Ne	O, Mg	Few years	One billion
O	Si, S	One year	Two billion
Si	Fe, Ni	Few days	Three billion

Elements are indicated by their chemical symbols, and billion means one thousand million

over yet again. Through these successive stages the star will produce oxygen, silicon, and finally iron. Each new nuclear reaction will only occur efficiently in a denser and therefore smaller (more interior) portion of the core than the previous one, and will also last for a shorter period of time. Thus the stellar interior will look not too different from an onion, whose various layers document the successive stages of the star's evolution. Table 6.3 shows some typical time scales for a star with 15 solar masses.

Nevertheless, this process does not continue repeating itself forever. When the core of the star turns into iron, all nuclear reactions stop permanently. The reason for this is that iron has the most stable nucleus of all the chemical elements, and all nuclear processes involving it are endothermic: they require more energy than they will produce. At this point the star has therefore run out of nuclear energy sources in the core that can balance its gravitational attraction, and it will therefore unavoidably collapse further.

Meanwhile, additional relevant physics is now happening at the subatomic level. The electrons are stripped off from the atoms and squashed closely together. It is at this point that quantum mechanical effects take over when it comes to determining what happens next—in short, how the star will die. It turns out that there are three possible outcomes, depending on the mass of the star. But before discussing them we need to look in more detail at the star's energy production mechanisms.

Poets say science takes away from the beauty of the stars—mere globs of gas atoms. I, too, can see the stars on a desert night, and feel them. But do I see less or more?

Richard Feynman (1918–1988)

6.2   Stellar Nucleosynthesis

While they are on the Main Sequence, stars obtain their energy by converting hydrogen into helium, producing both electromagnetic radiation and another type of particles called neutrinos. For example, the Sun is burning around 700 million tons of hydrogen every second, and its chemical composition is about 92.1% hydrogen, 7.8% helium, and 0.1% everything else. As discussed in the previous section, in the latter stages of their evolution stars can successively burn and produce heavier elements: helium can be burned into carbon, and similarly oxygen, neon, magnesium, silicon or iron may be produced, depending on the mass of the star in question. Through this process, stars play a crucial role in the formation of some of the elements in the periodic table—a process generally referred to as nucleosynthesis.

In fact, elements in different parts of the periodic table were produced at different times and in different places throughout the Universe. Most of the hydrogen and helium (and also a small amount of lithium) were produced in the early Universe, when its age was roughly between 1 and 3 min old. This is usually known as Big Bang nucleosynthesis (or primordial nucleosynthesis), and we will mention it again in the next chapter. The intermediate elements— up to the previously mentioned iron—were first synthesised in the cores of stars, and then disseminated through the interstellar medium once these stars died—we will see how in the next section. This is usually referred to as stellar nucleosynthesis. Finally, elements heavier than iron, such as silver, gold, or uranium, are only produced in specific and comparatively uncommon events, such as supernova explosions and violent mergers of two neutron stars. Naturally, such events also produce some of the lighter elements. By comparison to the other two processes, this last one is sometimes called explosive nucleosynthesis.

Note that among the products of stellar nucleosynthesis are elements such as the carbon that provides our energy, the oxygen we breathe, the calcium in our bones, and the iron in our blood. As has been often said, we are made of stellar ashes. Specifically, what this means is that the fact that we are carbon-based life forms implies that the Sun can't be a first-generation star. Instead, one or more stars had to be born, live for millions or billions of years, and then die in our galactic neighbourhood, thus disseminating these elements in the region where the Sun formed, so that they could be abundant in the Solar System.

While all Main Sequence stars are converting hydrogen into helium, not all of them do it in the same way. There are in fact two different processes, known

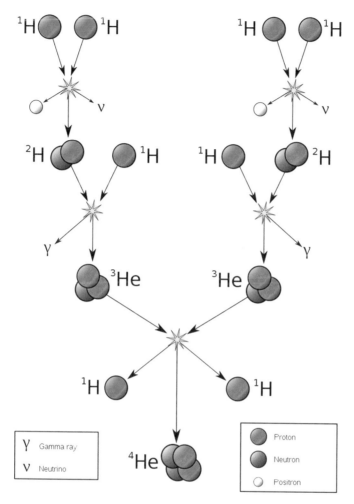

Fig. 6.3 Schematic view of the simplest of the proton–proton chain reactions, allowing stars to convert hydrogen into helium (Public domain image)

as the proton–proton chain reactions and the CNO cycle. Which one is the main source of the star's energy depends primarily on its mass.

In the proton–proton chain reactions—strictly speaking there are several of them, the simplest of which is depicted in Fig. 6.3—protons first combine to produce deuterium (a heavy isotope of hydrogen, whose nucleus contains one proton and one neutron). The subsequent addition of one proton forms the helium-3 isotope, and finally two helium-3 nuclei interact to produce a helium-4 nucleus, as well as two leftover protons. The overall process also produces positrons (the antiparticles of electrons), neutrinos, and photons. So by analogy with chemical reactions we could summarise the overall result of the process by writing

4 × hydrogen ⟶ helium + 2 positrons + 2 neutrinos + 2 photons

On the other hand, the CNO cycle is a cycle of reactions which—as the name suggests—involves several isotopes of carbon, nitrogen, and oxygen. (Again, strictly speaking there are several of these cycles, some of which also involve isotopes of other elements.) Although the nuclear reactions involved in this case are different and somewhat more elaborate, their overall balance turns out to be fairly similar. In the simplest cycle one has

4 × hydrogen ⟶ helium + 2 positrons + 2 neutrinos + 3 photons

The relative contribution of the two processes to the star's energy production budget depends on the star's mass, because the CNO cycle is only efficient at higher core temperatures and densities. For the case of the Sun, about ninety percent of its energy comes from the proton–proton chains and only ten percent from the CNO cycle. For stars with masses thirty percent greater than that of the Sun, the two processes make equal contributions to energy production, and for heavier stars, the CNO cycle is the dominant one.

These different energy production mechanisms also impact the structure of the star's interior, and in particular how the energy that is produced in the star's core is subsequently transported to the outside. There are in principle three possible transport mechanisms, whose efficiency will depend on various other physical properties.

Radiation (that is, the absorption and re-emission of photons); its efficiency decreases when the opacity, the density, or the photon flux increase.

Convection (hot regions rise, expand, and cool); its efficiency depends on temperature gradients, that is the temperature differences between lower and upper layers of the star.

Conduction (collisions between electrons); its efficiency also depends on the temperature gradient.

Conduction is the easiest one to deal with, because it is always subdominant in Main Sequence stars, although it does play an important role in the internal structure of white dwarfs.

In stars like the Sun, the temperature of the outer layers is not sufficient to ionise hydrogen, so photons coming up from the interior are very easily absorbed, meaning that the opacity is high. Thus the outer layers of the Sun (and those of other low-mass stars) are convective. On the other hand, in heavier stars the temperature of the outer layers will be higher and the hydrogen there will be ionised, so the opacity is much lower and radiative transport will be the dominant mechanism.

How about the core of the star, where nuclear reactions are taking place and the star's energy is being produced? If the energy production mechanism depends strongly on temperature, then there will be a steep gradient between the core's layers (remember that the temperature and density will be highest at the very centre of the core and will decrease as one moves away from it). In that case the radiative flux will be high and radiative transport will be insufficient to pump all the energy outwards. For low-mass stars (including the Sun), for which the proton–proton chains dominate energy production, the amount of energy produced depends on the fourth power of the temperature; this turns out to be a comparatively weak dependence, and in this case the core will be radiative. On the other hand, for heavier stars, for which the CNO cycle dominates, the energy produced depends approximately on the seventeenth power of the temperature. This leads to very steep temperature gradients and therefore a convective core.

Thus we end up with a very interesting situation: low-mass stars have a radiative core and convective outer layers, while high-mass stars have a convective core and radiative outer layers. In other words, a small star is not simply a miniature version of a larger star! This nicely illustrates how the same two competing physical mechanisms can lead to rather different outcomes in the same kind of system when other relevant quantities are changed.

6.3 White Dwarfs, Neutron Stars, and Black Holes

Having seen how stars form and live, we can now discuss how they die. As has already been mentioned there are three possible outcomes, depending on the star mass, and each with it own peculiarities and interesting features.

The first scenario occurs for stars whose cores are less than about 1.4 times the mass of the Sun. The existence of such a threshold was deduced theoretically by Subrahmanyan Chandrasekhar (1910–1995), so the corresponding mass is known as the Chandrasekhar mass limit. Notice that this applies to the core of the star. The total mass of the progenitor star can be up to ten solar masses, which encompasses about ninety-eight percent of the stars in our galaxy.

In this case, it turns out that a quantum effect called electron degeneracy pressure will be sufficient to stop the gravitational collapse of the star's core. This pressure is the result of Pauli's exclusion principle, which expresses the fact that particles such as electrons don't like to be squeezed too close together (while others such as photons do not mind), and will exert a repulsive pressure if one tries to do so. Note that, although electrons will of course repel one another electromagnetically, this is a different physical effect. The Chandrasekhar mass is the value beyond which the gravitational attraction becomes too strong even for the electron degeneracy pressure to be able to balance it.

Thus the core collapse is stopped, but the core itself contains only part of the star's mass, and its gravitational field is not strong enough to hold onto the outer layers of the progenitor star; these therefore are able to escape and gradually drift away. The final result is an expanding cloud of stellar material, at the centre of which remains a small bright star-like object. The first of these is what astronomers call a planetary nebula—an old term which was possibly first used by William Herschel (the discoverer of Uranus) and reflects both the fact that they often looked like planets when seen in comparatively small telescopes and the fact that they were initially interpreted as being clouds containing a young star where planets were still forming.

The second of these, the remainder of the core, is called a white dwarf. Although the electron degeneracy pressure manages to stop the collapse, it can only do it once the electrons are sufficiently close together. As it turns out, the core size has to be reduced to that of a rocky planet such as the Earth. And indeed, the larger the white dwarf's mass the smaller its radius will be—which can easily be understood bearing in mind that a larger mass means a stronger gravitational pull, which forces the electrons to squeeze together more tightly. In fact, one can show that as the mass approaches the Chandrashekhar limit the radius approaches zero, highlighting the fact that the electron degeneracy pressure becomes incapable of preventing the collapse of such an object. In any case, this also implies that white dwarfs are extremely dense: the usual factoid is that one teaspoon of material on it would (nominally) weigh 1 ton on Earth.

Typical white dwarfs in our galaxy tend to have masses around sixty percent of the Sun's mass and a radius of about one percent of the Sun's radius—again, about the same as the Earth's radius. They are also quite hot, with temperatures ranging from a few thousand to a few tens of thousands of degrees. But very slowly they radiate this energy away, thereby cooling down. Eventually one expects them to lose all their energy, or at least to reach the same temperature as the cosmic microwave background (which we will discuss in the next chapter), becoming what have been dubbed black dwarfs. Since white dwarfs typically have a substantial amount of carbon (as well as some oxygen and neon), one may think of them as Earth-sized lumps of coal. However, the time scale necessary for such black dwarfs to form is thought to be many orders of magnitude longer than the current age of the Universe, so we do not expect any of them to exist at the moment anywhere in the Universe.

A star whose iron core mass is greater than the Chandrasekhar limit of 1.4 solar masses but below 3–4 solar masses (the exact number is not well known in this case) will have a much more spectacular end. In this case, this corresponds to a total mass of the progenitor in the range between 10 and about 29 solar masses. Here the electron degeneracy pressure is unable to avoid the star's sudden collapse—on a timescale of not more than minutes—from an original size significantly larger than that of the Sun, to that of a city. The result is what astronomers call a supernova explosion—technically astronomers distinguish several types of supernova explosions, this one being known as a type II supernova.

In the process of collapse the nuclei of iron and other elements are broken into neutrons and protons, and these protons combine with electrons to produce more neutrons, as well as neutrinos. When the core of the former star reaches the size of a dozen kilometres or so, the neutrons are being crushed so close together that their degeneracy pressure becomes significant and together with other repulsive nuclear forces they are once again able to interrupt the collapse. The leftover object is called a neutron star. Incidentally, here you can also see that this degeneracy pressure is not an electromagnetic effect, since neutrons have zero electric charge.

This sudden break produces a gigantic shock wave that blows away the star's outer layers. The energy released in the process temporarily enables a series of previously unfeasible nuclear reactions that produce elements heavier than iron—the explosive nucleosynthesis mentioned in the previous section. The explosion also scatters these heavy elements, together with the lighter ones previously formed inside the star, through the surrounding interstellar medium. Note that the ejected material adds up to many solar masses. This shock wave can itself trigger the collapse of some neighbouring gas clouds, thereby starting the process of formation of the next generation of stars.

Remarkably, only a fraction of the amount of energy released during the contraction is in the form of visible light, or of photons of any wavelength. Most of the energy is actually carried away by the neutrinos, which leave the exploding star unopposed. Thus although supernova explosions are already extremely bright as seen through the photons they emit, they would be even brighter if we were able to see all the emitted neutrinos.

The remnant neutron star will be even denser and hotter than a white dwarf. Their masses are though to range between the lower 1.4 solar mass Chandrasekhar limit and an upper limit of about 2.1 solar masses which is known as the Tolman–Oppenheimer–Volkoff limit. The typical radius of neutron stars is between 10 and 20 km, and as in the case of white dwarfs a larger mass usually corresponds to a smaller radius, although in this case the relation between the two is less certain because it is expected to depend on the internal structure and composition of neutron stars—which currently is not well known at all.

Just as the initial gas cloud that led to the star gradually increased its rate of rotation as it contracted, so the star's core and the resulting neutron star will speed up as it reduces it size, typically reaching a rotation rate of around 30 times per second—a remarkable number once you recall that we're talking about an object more massive than the Sun. Although this is a typical number, there is a broad range of rotation frequencies: at the time of writing the fastest known has a frequency of 716 rotations per second, while the slowest one only rotates once every 23.5 s.

The collapse also concentrates the magnetic field (which all stars have, to a larger or smaller extent), so we end up with a very intense and collimated beam of X-rays that rotates with the star—think of an X-ray lighthouse. There is no reason why the neutron star's rotation axis and that of its magnetic field need to be aligned. When, by chance, the X-ray beam happens to point towards the Earth we detect an extremely regular X-ray pulse. This type of neutron star is called a pulsar—as in several other cases in astronomy, the name was introduced before the physics of these objects was fully understood. The first pulsar was discovered in 1967 by Jocelyn Bell Burnell (b. 1943) and Antony Hewish (b. 1924), and several thousands are now known, including various interesting examples such as a double binary pulsar—a binary system where both objects are pulsars.

Pulsars (see Fig. 6.4 for an example) have several remarkable properties. Their surface gravitational field is extremely strong, as a result of a density whose typical value is around 600 million metric tons per cubic centimetre.

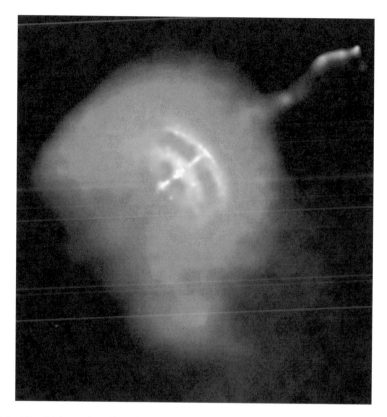

Fig. 6.4 The Vela pulsar (the bright white spot in the image), together with its collimated jet and surrounding pulsar wind nebula. Credit: NASA/CXC/PSU/G.Pavlov et al.

This means that freely falling objects will hit its surface with a speed of about half the speed of light. As we saw when discussing general relativity, the gravitational field affects how clocks run, and on the surface of a neutron star clocks will also run at about half the speed of those on the Earth. And the local curvature is such that light is very significantly bent: if you were sitting on the surface of a neutron star you would be able to see 20 or 30° beyond the horizon.

Their magnetic fields are also extremely strong. A typical value is about ten million times larger than that of the most powerful magnetic resonance imaging devices—and these are in turn about one thousand times stronger that of a typical fridge magnet, and more than one hundred thousand times stronger than the Earth's magnetic field. And there is a particular class of pulsars, known as magnetars, that has even stronger magnetic fields. Finally, as a curiosity, the first known exoplanets were discovered by Aleksander Wolszczan (b. 1946) and Dale Frail (b. 1961) in 1992 orbiting a pulsar, which is now known to have three planets (or possibly two planets and one asteroid).

As time goes by, the neutron star's energy is slowly radiated away, and the rotation will slow down correspondingly. Thus the broad range of rotation speeds reflects in part the ages of the different pulsars, although other factors are also relevant. Eventually, so much energy will be lost that the neutron star is no longer able to generate the X-ray beam. When all the energy is spent the star again ends up as a very dense but dark cinder. Although the rotation slowdown can be very accurately measured—and is in excellent agreement with what is predicted by general relativity—the timescale for neutron stars to cool down completely is, as in the case of white dwarfs, many orders of magnitude longer than the current age of the universe.

Finally, for stars whose core is heavier than about 3–4 solar masses (so progenitor masses larger than about 29 solar masses), or whose putative neutron stars would be heavier than the 2.1 solar mass limit, the outcome of the collapse is even more dramatic. As for the intermediate mass stars, the core will collapse to planet size within a timescale of minutes, and a supernova explosion follows, with all the astrophysical consequences that have already been discussed. The difference is that now there is no known physical mechanism that can stop the collapse of the core, which at least in an abstract way must continue indefinitely, leading to the formation of a black hole.

Since only the more massive stars explode as supernovas and heavy stars are comparatively uncommon, in a galaxy like ours there is an average of one supernova per century. The last two have in fact already been mentioned: they occurred in 1572 and in 1604, and are commonly known as Tycho's supernova and Kepler's supernova (see Figs. 6.5 and 6.6, respectively). It is interesting that, happening when they did, they played a significant historical role in the birth of modern science, as has been discussed earlier in the book.

Although the current supernova death rate of one per century in our galaxy is rather low compared with the star birth rate of seven per year, there are more than one hundred billion galaxies in the visible Universe. Therefore, and although different galaxies will have different supernova death rates (depending on their size, age, and other properties), one may estimate that somewhere in the visible Universe a star is dying as a supernova every second, thereby creating improved conditions for the emergence of life elsewhere.

The most recent supernova in our cosmic neighbourhood, which astronomers have been able to study in some detail, became known as SN

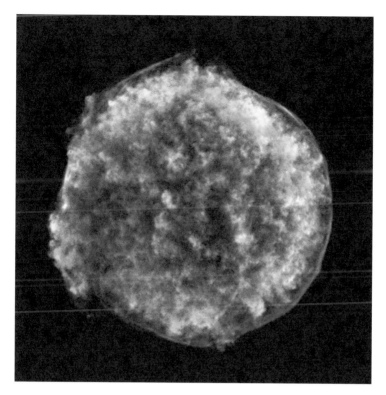

Fig. 6.5 Remnant of the 1572 supernova (known as Tycho's supernova) as seen by the Chandra X-ray Observatory. The red and green colors identify the expanding plasma cloud, which has temperatures of millions of degrees. Credit: NASA/CXC/Rutgers/J.Warren & J.Hughes et al.

1987A. It exploded in the Large Magellanic Cloud (a dwarf galaxy that is one of the satellites of our Milky Way galaxy), and was observed from the Earth on 23 February 1987, although given its distance the explosion actually occurred some 168,000 years ago. The progenitor star had an estimated twenty solar masses, so the remnant of the explosion was a neutron star, whose presence was observationally confirmed in 2019.

Actually, a total of about 25 neutrinos from this explosion were detected on Earth by two instruments. This is interesting for two separate reasons. The first reason is that neutrinos are extremely weakly interacting—in other words, their probability of interacting with ordinary matter is extremely small. The fact that they were detected at all therefore implies that extremely large numbers of them must have been emitted, which in turn highlights the

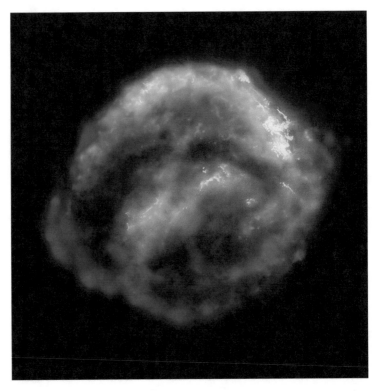

Fig. 6.6 A false-color composite image of the remnant of the 1604 supernova (known as Kepler's supernova) from pictures by the Spitzer Space Telescope, Hubble Space Telescope, and Chandra X-ray Observatory. Credit: NASA/ESA/JHU/R.Sankrit & W.Blair

uncommonly large amount of energy released in the explosion. And the second reason is that they were detected a few hours before the visible light from the explosion reached the Earth. This is because neutrinos can escape immediately during the core collapse, while visible light only escapes once the shock wave reaches the stellar surface.

Although the term 'black hole' is recent, having been introduced in the 1960s (with its originator being disputed), the fact that objects could have gravitational fields so strong that even light would not be able to escape them was already noticed in the eighteenth century. John Michell (1724–1793) suggested in a paper read at the Royal Society in 1783 that what he called 'dark stars' could exist, and Pierre-Simon de Laplace (1749–1827) also proposed a similar idea. Michell further pointed out that they could be identified by looking for astrophysical systems that behaved as if they were composed of two objects but only one of them could be seen:

> If there should really exist in Nature any bodies, whose density is not less than that of the sun, and whose diameters are more than 500 times the diameter of the sun, since their light could not arrive at us; or if there should exist any other bodies of a somewhat smaller size, which are not naturally luminous; of the existence of bodies under either of these circumstances, we could have no information from sight; yet, if any other luminous bodies should happen to revolve about them we might still perhaps from the motions of these revolving bodies infer the existence of the central ones with some degree of probability, as this might afford a clue to some of the apparent irregularities of the revolving bodies, which would not be easily explicable on any other hypothesis; but as the consequences of such a supposition are very obvious, and the consideration of them somewhat beside my present purpose, I shall not prosecute them any further.
>
> John Michell (1724–1793)

The fate of the core material once its surface crosses the black hole's event horizon is unknown, although it has of course been the subject of extensive mathematical study and speculation. The event horizon defines the region beyond which the escape of anything from the black hole is impossible. Formally, one expects the density to keep growing indefinitely, very quickly exceeding any density that has been experimentally probed, directly or indirectly, in physics laboratories. It's possible that the collapse will indeed continue indefinitely, but it is also conceivable that it is stopped by physical mechanisms, happening inside the event horizon, that are currently unknown to us.

For non-rotating black holes the size of the event horizon is known as the Schwarzschild radius, after Karl Schwarzschild (1873–1916) who was one of the first to study exact mathematical solutions of general relativity and specifically studied this case in 1916, shortly before dying on the Russian front during World War I. (For rotating black holes, the event horizon and the Schwarzschild radius become different concepts.) If you imagined squeezing the Sun until it became a black hole, you would need to squeeze all its mass into a size of about 3 km, while in the case of the Earth the Schwarzschild radius is only about 9 mm.

As far as we know there is no limit to the mass of astrophysical black holes, and there is strong evidence for the existence of black holes with millions or even billions of solar masses in the centre of various galaxies, including our own Milky Way. The recent detection of binary black hole mergers by the LIGO and Virgo collaborations and the even more recent imaging of a black

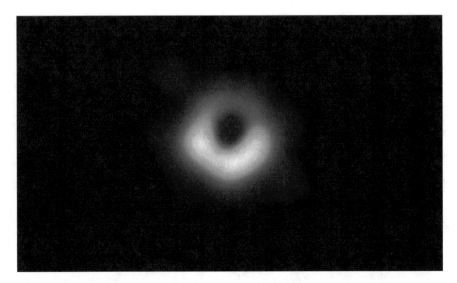

Fig. 6.7 The shadow of the supermassive black hole at the centre of Messier 87, imaged by the Event Horizon Telescope collaboration. Credit: EHT Collaboration

hole shadow by the Event Horizon Telescope collaboration (see Fig. 6.7) have significantly consolidated this evidence.

In addition to these astrophysical black holes, in many cosmological models one also finds black holes produced in the early Universe, known as primordial black holes. These can have much smaller masses, and among other things have been conjectured to provide at least some of the Universe's dark matter (which we will discuss in the following chapter). Just like white dwarfs and neutron stars, black holes are also not entirely stable or permanent objects. Instead they evaporate, at an extremely slow rate, through a quantum mechanical process first identified by Stephen Hawking (1942–2018) and known as Hawking radiation.

6.4   Galaxies and the Expanding Universe

If you go stargazing on a clear and moonless night and in a place without light pollution, you will soon notice that you can see more than just stars and planets. Even with the naked eye alone you will see a few fuzzy or cloudy patches of light that are clearly different from stars. Of course, early astronomers noticed them too, and not quite knowing what they were looking at simply called them—in modern English—nebulas, meaning clouds.

Binoculars or telescopes reveal many more of these nebulas, and nice as they are they have at least one disadvantage: if you happen to be looking for comets, you may confuse them with one. Thus in the eighteenth century Charles Messier (1730–1817), who indeed was mainly interested in comets, put together a catalogue—that now bears his name—of more than one hundred such fuzzy objects in the northern hemisphere of the sky that might be mistaken for comets.

As observational facilities improved and more of these nebulas were identified and catalogued, it gradually became clear that they actually have a variety of somewhat different forms. Some seemed to be just gas clouds, others were groups of stars closely packed together, still others were clearly spiral-shaped. This naturally led to speculations about their possibly different origins and properties, including the most basic ones such as their distance.

William Herschel also became interested in the spatial distribution of the stars around us. Counting the numbers of stars in different directions of the sky he found that most of them were in a flattened disk structure across the sky, with the numbers of stars being approximately the same in all directions around this structure. In 1785 he made a map (see Fig. 6.8), summarizing his

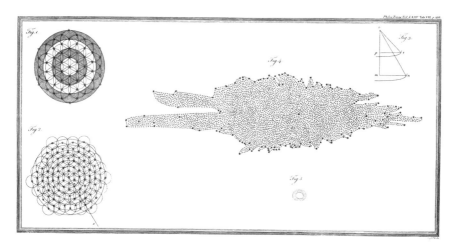

Fig. 6.8 William Herschel's map of the Milky Way, published in 1785 (Public domain image)

results and concluded that the Sun was approximately at the centre of this disk. In his own words:

> That the Milky Way is a most extensive stratum of stars of various sizes admits no longer of the least doubt and our sun is actually one of the heavenly bodies belonging to it as evident. I have now viewed this shining zone in almost every direction, and find it composed of stars, whose number, by the account of these gages, constantly increases and decreases in proportion to its apparent brightness to the naked eye.
>
> William Herschel (1738–1822)

By the beginning of the twentieth century there was robust evidence that the Solar System was indeed part of a very large system containing billions of stars as well as gas and dust: a galaxy, known by the classical Greek name 'Milky Way', which originally referred to its densest and most luminous part. This is the part that one can see with the naked eye on a clear and moonless night. But where do all the nebulas fit in? While most of them seemed to be relatively nearby (and were plausibly part of the Milky Way), the situation was far less clear in the particular case of the spiral ones. They could be found on the sky in all directions away from the Milky Way, but were they part of it or were they perhaps galaxies (or island universes, in the terminology of the time) of their own?

Answering the question required knowing two things. The first was: how large is the Milky Way, in other words how far out does it extend? And the second was: how far away are the spiral nebulas, which would also reveal their true size (the angular size on the sky being of course known). In other words, the problem was one of determining two distances, and distance determination is a notoriously difficult problem in astronomy.

Meanwhile, Vesto Slipher (1875–1969), initially hired by the Lowell Observatory (in Flagstaff, Arizona) to study Mars and other Solar System objects, became interested in the spiral nebulas, and in particular formulated the plan of determining whether they were new stellar systems in the process of formation. His envisaged method was to systematically obtain the spectra of these objects in order to check whether the chemical elements found at the edges of the spiral nebulas matched the chemical composition of the giant planets in the outer part of the Solar System.

Taking the spectrum of Andromeda—or M31 in Messier's catalogue, and incidentally, the most distant object that you can see with the naked eye, being 2.5 million light years away—at the end of 1912 he found that it was moving towards us at a speed of 300 km/s. This was a somewhat unexpected value, being significantly larger than the average speed of stars in our own galaxy. Despite his reasonable doubts, his measurement was remarkably accurate: now we know that the correct speed is 301 ± 1 km/s.

Gradually, additional measurements increased his confidence in his spectrograph and his observational techniques, but also revealed an intriguing pattern: most of these objects had redshifted spectra, meaning that they were moving away from us, and moreover the fainter ones tended to have larger recession speeds. In the summer of 1914 he had accumulated measurements for 15 nebulas, of which 12 were redshifted and only 3 blueshifted (one of them being Andromeda). Three years later he had increased his dataset to 25 nebulas, of which 21 were redshifted, and by 1925 he had reached 45, and the trend remained, strongly suggesting that these objects must be extragalactic. However, for conclusive evidence, it was still necessary to find a way to determine the distance to these objects.

Not all stars shine with a constant luminosity. A significant fraction of them are known as variable stars, since their intrinsic luminosity changes over time, in irregular or regular ways as the case may be. Various physical mechanisms can make the star expand and contract at regular intervals, oscillating about an equilibrium size. Two classes of these regular (or pulsating) variable stars turned out to be particularly useful for determining (comparatively) nearby distances in astronomy. The first are known as RR Lyrae (so named after the brightest star in this class), and they always have the same pulsation period and intrinsic luminosity. The first of these properties allows astronomers to identify them as belonging to this class, while the second implies that by measuring their apparent luminosity one can immediately infer their distance.

Around 1917 Harlow Shapley (1885–1972) used RR Lyrae stars in globular clusters to measure their distances. These globular clusters were one of the classes of nebulas, some of them being part of Messier's catalogue. They are spherically symmetric collections of hundreds of thousands of densely packed and fairly old stars. By studying almost one hundred of these (we now know about 150 within the Milky Way), he found that they had an almost spherical spatial distribution, but the centre of this distribution was not anywhere near the Solar System. Instead, it was rather far away, towards the direction of the constellation Sagittarius. He thus hypothesised that this centre was also the centre of the Milky Way, which we now know to be correct. We also know that the galactic centre is about 26,700 light-years away, and hosts a supermassive black hole with about 4.1 million solar masses.

This also shows that the Sun (and the Solar System) do not occupy a special place within the Milky Way. In fact, and like all the other stars in it, we are

Fig. 6.9 Henrietta Swan Leavitt, photo date unknown (Public domain image)

orbiting it (in an almost circular orbit) with an orbital speed of about 230 km/s. The corresponding orbital period, which is estimated to be between 225 and 250 million years, is often called the galactic year or sometimes the cosmic year. Thus on this scale the Earth is no older than about 20 galactic years, and in all the time that humans have been looking at the night sky the appearance of the Milky Way (which, as has been pointed out, you can see with your naked eye) has only changed by a tiny amount.

The second and more interesting class of pulsating stars are the Cepheids, named after Delta Cephei, the first star of this type to be identified as a variable star—by John Goodricke (1764–1786) in 1784. In this case they can have different pulsation periods as well as different intrinsic luminosities, but in 1908 Henrietta Leavitt (1868–1921, see Fig. 6.9) discovered that the two are correlated. Larger and brighter stars will pulsate more slowly (that is, with a longer period) than smaller and fainter ones.

The issue would be resolved through the 1920s by Edwin Hubble (1889–1953), with the often forgotten but in fact essential help of Milton Humason (1891–1972), benefiting from the recently inaugurated 100-inch Hooker telescope at the Mount Wilson Observatory. By 1924 Cepheid variables could be identified in several spiral nebulas (again including Andromeda), and this enabled a determination of their distances. These conclusively implied that they were well beyond our galaxy, whose size was by then reasonably established. Thus spiral nebulas are indeed galaxies like our own Milky Way.

Astronomers define the redshift as the relative change in the observed wavelength of a spectral line, by comparison to the emitted wavelength, that is,

$$z = \frac{\lambda_{obs} - \lambda_{em}}{\lambda_{em}}. \tag{6.1}$$

A blueshift, as in the case of the Andromeda galaxy and a few of our other neighbours, would therefore correspond to a negative redshift. If the redshift is interpreted as being due to the motion of the galaxies it follows that they are moving away from us, and Slipher's results further suggested that the fainter ones are typically doing it faster.

Note that for the case of small velocities (and correspondingly small redshifts) the two quantities are simply related by

$$z = \frac{v}{c}, \tag{6.2}$$

which you might recognise as being mathematically the same as the formula for the Doppler effect. However, as we will see in the next chapter, the physical interpretation is different here. In the case of the Doppler effect the underlying physics is a relative speed between the source of the observer, which in our case would correspond to the galaxies moving in a fixed pre-existing background. This corresponds to the Newtonian view, which as has already been discussed in the previous chapter is not correct. In our case it is space itself that is being stretched; apart from possible local peculiar velocities (to be discussed in a moment), the galaxies are not moving.

A further limitation of that formula is that it can't possibly apply to even moderately large redshifts, since it would imply a recession speed faster than that of light. Indeed, the correct relation, which is easily obtained in special relativity, is

$$\frac{v}{c} = \frac{(1+z)^2 - 1}{(1+z)^2 + 1}, \tag{6.3}$$

which as you can easily see never exceeds the speed of light. Alternatively, we can invert this expression and write the redshift as a function of the recession speed as follows:

$$1 + z = \sqrt{\frac{1 + \frac{v}{c}}{1 - \frac{v}{c}}} . \tag{6.4}$$

If you want you can check that both of these reduce to the previous Eq. (6.2) in the limit of small speeds and redshifts. So this is analogous to the relation between the Newton and Einstein equations: the former can be recovered as a particular mathematical limit of the latter, but their physical contents and interpretations are clearly different.

Getting back to the 1920s, there was growing evidence that the recession speeds increased with distance, but what was the specific relation between the two? Early attempts by Carl Wirtz (1876–1939) between 1922 and 1924 and by Knut Lundmark (1889–1958) also in 1924 were inconclusive due to the limited data available to them. By the end of the decade, and building on Slipher's work, Hubble and Humason had collected distance and redshift data of enough galaxies to claim the empirical relation

$$v = H_0 d , \tag{6.5}$$

with the proportionality factor known as the Hubble constant. The law itself is now known as the Hubble–Lemaître law, since Georges Lemaître (1894–1966) had already predicted theoretically in 1927, based on general relativity, that a redshift–distance relation should exist, and indeed he also plotted the speed versus distance data available to him and obtained an estimate of the Hubble constant that was not very different from the one found by Hubble himself.

Thus the Universe is expanding and at intermediate distances the galaxies are moving away from us according to the Hubble–Lemaître law, with a recession speed that is roughly proportional to their distance. At small distances the law need not apply to all the galaxies, since the gravitational fields due to the mutual attraction of nearby galaxies can overcome the effect of the overall expansion, leading to the so-called peculiar velocities. This is the reason why the Milky Way and the Andromeda galaxy are moving towards each other (as first measured by Slipher), and the two will collide (in a broad physical sense) in about four billion years (see Fig. 6.10).

Fig. 6.10 A simulation, based on data from the Hubble Space Telescope, of the Earth's night sky in about 3.75 billion years and the Andromeda galaxy (on the left) and the Milky Way. Credit: NASA

On the other hand, at very large (cosmological) distances the relation will depend on the contents of the Universe. Moreover, in this case there is a second effect: the expansion rate of the Universe is not a constant but actually depends on the redshift, so one has a redshift-dependent Hubble parameter rather than simply a Hubble constant. We will discuss this further in the next chapter.

7

The Physics of the Universe

In this chapter we review our current view of the Universe, which is contained in what is called the Hot Big Bang model. We discuss its main assumptions and successes, but also its shortcomings. The most glaring of these is the fact that most of the contents of the Universe are in the form of two components, known as dark matter and dark energy, which so far have not been directly detected and are only known indirectly from various astrophysical observations. We will also mention some of the many directions that have been explored in order to extend this model, involving scalar fields and extra dimensions.

> Cosmology is peculiar among the sciences for it is both the oldest and the youngest. From the dawn of civilization man has speculated about the nature of the starry heavens and the origin of the world, but only in the present [twentieth] century has physical cosmology split away from general philosophy to become an independent discipline.
>
> Gerald Whitrow (1912–2000)

7.1   The Hot Big Bang Model

In the broad sense, every culture has made an effort to understand its place in the Universe. This effort naturally leads to the development of cosmological models, which are descriptions (possibly extremely simplified ones) of reality. Depending on various circumstances these descriptions may be based on

© Springer Nature Switzerland AG 2020
C. Martins, *The Universe Today*, Astronomers' Universe,
https://doi.org/10.1007/978-3-030-49632-6_7

mythological, religious, philosophical, or scientific principles. From this broad point of view, cosmology could arguably be described as the oldest of sciences. As we have seen, the first broadly accepted cosmological model, that of Aristotle and Ptolemy, developed from about 2500 years ago, and was the standard cosmological model for about 2000 years.

Despite great differences in the details of various models (some of which have been mentioned in the first chapters of this book), most early cosmological models have two important things in common. The first is that they are static models: at least on sufficiently large scales, the structure of the Universe is always the same and does not evolve with time. This is no longer the case: one of the most fundamental aspects of our current view of the Universe is that it evolves in time. In particular, it is expanding, as we saw at the end of the last chapter.

A second common aspect is the separation between the celestial and terrestrial parts of the model. Although we have known for several centuries that there is no real difference, in the sense that the same laws apply in both cases, a relic of this separation still exists today, in the division between physics (which at least implicitly describes the Universe around us) and astronomy (which describes the Universe outside the Earth). This is more than a question of semantics. For example, many universities simultaneously offer degrees in both topics. Such a division is artificial, as both disciplines are needed if we want to understand the origin and evolution of the Universe—which is the main goal of cosmology.

Indeed, the early Universe is above all a privileged laboratory where many key aspects of fundamental physics can be tested with an accuracy that is not achievable in local laboratories or particle accelerators. Our standard cosmological model, known as the Hot Big Bang model, gradually developed and gained acceptance during the twentieth century. Crucially, about two decades ago cosmology underwent a transformation. Up to that point, despite some essential experimental inputs that occasionally came in, it had been a science open to significant amounts of theoretical speculation, but in the space of a few years it become strongly driven by observational results. Indeed, we have learned more about the origin and evolution of the Universe in the last 20 years than in the rest of humanity's history.

Perhaps the most significant of the many surprises that these observational efforts have revealed is the fact that about 96% of the contents of the Universe is in an unknown form—or, to be more rigorous, two unknown forms—that so far we have never directly detected in laboratory experiments, but are inferred by statistical analyses. In other words, we are seemingly discovering these components mathematically (much in the same way as Neptune was initially discovered mathematically), but so far we have not discovered them physically.

But a second revolution is already taking place: this decade will see the emergence of a new generation of instruments that will allow astrophysicists and cosmologists to probe the fundamental physical laws of the Universe with unprecedented accuracy. We will discuss some examples of these tests, and of how this new approach can be crucial to the resolution of key enigmas in modern cosmology and fundamental physics, in the next chapter. For the moment we will focus on the current status of the Big Bang model.

Cosmology studies the origin and evolution of the Universe, and in particular of its large-scale structures, on the basis of physical laws. By large-scale structures I mean scales of galaxies and beyond. This is the scale where interesting dynamics is happening today: anything happening on smaller scales is largely irrelevant for cosmological dynamics. The Hot Big Bang model was developed from the 1920s, starting with the mathematical work of Alexander Friedmann (1888–1925, see Fig. 7.1) and Georges Lemaître (1894–1966) and became almost universally accepted from 1965. It provides an observationally tested description of the evolution of the Universe from a time one hundredth of a second after the beginning until the present day. On the other hand, it leaves some questions unanswered (in particular about its initial conditions), which means that it is an incomplete model and is therefore destined to be replaced eventually by a better one.

The model rests on three conceptual pillars, which are worth understanding:

The first is called the cosmological principle, which states that on sufficiently large scales the Universe is homogeneous (there is no preferred place) and isotropic (there is no preferred direction).

The second is the assumption that Einstein's general relativity, which we discussed previously, is the correct physical theory to describe the dynamics of the Universe.

The third is the hypothesis that the contents of the Universe can be described as one or more components each of which is a perfect classical gas (or Maxwell gas), which in a nutshell means that each of the Universe's contents has a pressure proportional to the density.

It is worthy of note that the first of these stems from the empirical observation that whenever we thought that we occupied a special place in the Universe, this turned out not to be the case. None of the Earth, the Sun (or Solar System) or our own galaxy are the centre of the Universe, as we once thought.

Fig. 7.1 Alexander Friedmann (Public domain image)

This therefore suggests that our large-scale view of the Universe should be a typical one. In other words, and at least in a statistical sense, we should see the same large-scale properties of the Universe as any other observer. This physical assumption is then expressed mathematically as the assumption of homogeneity and isotropy. As for the third assumption, the operative word is 'classical' (as opposed to 'quantum'): for most of its evolution the density of the Universe has been sufficiently low that explicitly taking into account the quantum nature of the particles is, to a very good approximation, not necessary. It only becomes important to do it in the very early Universe.

The main result of these assumptions is summarised, in cartoon form, in Fig. 7.2. The Universe started out about 13.8 billion years ago, in an extremely hot and dense state, and it subsequently evolved, expanding and cooling down

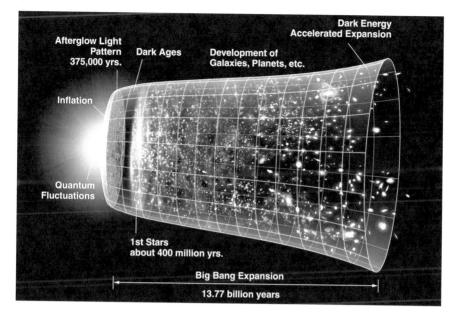

Fig. 7.2 Cartoon version of the evolution of the Universe, as described by the Hot Big Bang model. Credit: NASA/WMAP Science Team

as it did so. Although it was initially homogeneous and isotropic, some initial density fluctuations appeared at some point (to be discussed shortly), and these were then gravitationally amplified until they led to the large scale-structures—the galaxies and galaxy clusters—that we observed today. This gravitational amplification mechanism is essentially the same that has already been described for the formation of stars, only now applying on a grander scale. An overdense region will have a comparatively strong gravitational field and will therefore attract neighbouring matter, which will increase its overdensity, enabling it to attract even more matter, and so forth.

More specifically, the evolution of the Universe can be divided into six different phases, which are listed in Table 7.1. Three of these phases—those that are highlighted in bold on the table—are well characterised and their existence and main properties are supported by considerable observational evidence. Conversely, the other three phases are less well known, although significant efforts are ongoing, both on the theoretical and (especially) on the observational fronts, with the aim of understanding them better and searching for their more direct fingerprints.

Table 7.1 The six main phases of the evolution of the Universe, and some of their possible (expected) or actual observables

Phase	Observable(s)
Primordial era	Varying fundamental constants? Superstrings?
Inflation era	Primordial density fluctuations Gravitational waves?
Thermalization era	Dark matter?
Radiation era	Primordial nucleosynthesis Topological defects?
Matter era	Cosmic microwave background Galaxies and quasars
Acceleration era	Dark energy? Equivalence principle violations?

The two best known epochs in the evolution of the Universe are known as the radiation and matter eras. The former is the epoch whose dynamics was controlled by relativistic particles (such as photons), while the latter is the epoch whose dynamics was controlled by non-relativistic particles (such as ordinary matter). Observational probes of the radiation era include the primordial abundances of light elements, while those of the matter era include the large-scale structures such as galaxies and quasars. Another important probe is the cosmic microwave background, which formed in the matter era but not too long after the transition from the radiation era, and hence encodes a plethora of information on different epochs.

Strong evidence suggests that before these two epochs there was an inflation era. This is characterised by a period of extremely fast (possibly exponential or at least nearly exponential) expansion. This is peculiar because one can show that a Universe dominated by radiation or matter can never expand that fast. More specifically, what this ultimately implies is that the component dominating this phase of the Universe must have been mathematically very similar to a cosmological constant. Interestingly, Einstein had introduced such a cosmological constant earlier on, in a different historical context: expecting the Universe to be static, but finding that general relativity typically predicted an expanding or contracting Universe, he introduced this constant term (which was logically possible) in his equations to try to enable static solutions. Physically, this cosmological constant corresponds to a vacuum energy density (to which we will return later in the chapter), the relation between the two having been clarified by Yakov Zeldovich (1914–1987).

The theoretical view of the actual universe, if it is in correspondence to our reasoning, is the following. The curvature of space is variable in time and place, according to the distribution of matter, but we may roughly approximate it by means of a spherical space. [...] this view is logically consistent, and from the standpoint of the general theory of relativity lies nearest at hand [i.e., is most obvious]; whether, from the standpoint of present astronomical knowledge, it is tenable, will not be discussed here. In order to arrive at this consistent view, we admittedly had to introduce an extension of the field equations of gravitation, which is not justified by our actual knowledge of gravitation. It is to be emphasised, however, that a positive curvature of space is given by our results, even if the supplementary term [cosmological constant] is not introduced. The term is necessary only for the purpose of making possible a quasi-static distribution of matter, as required by the fact of the small velocity of the stars.

Albert Einstein (1879–1955)

It is in this inflation phase that we believe that the primordial density fluctuations that eventually led to the Universe's large-scale structures originated. They were simply quantum fluctuations (intrinsic to any quantum system) which, although originally of microscopic size, were stretched to macroscopic sizes due to the exponential expansion of the Universe, and subsequently grew due to the already mentioned gravitational amplification mechanism. This leads to one basic requirement for this inflation phase: it must last long enough (thereby providing a sufficient amount of stretching) to allow quantum fluctuations to grow to cosmological sizes.

As for the other three phases, very little is know about the primordial phase, in part because the inflation phase almost completely erases (or, more rigorously, exponentially dilutes) any possible relics of the earlier phase. However, in some cases it may be possible to infer some of the properties of this phase indirectly. At present these are quite poorly constrained, but on the positive side this means that it is possible that this phase will be found to contain significant clues to new physics beyond the current standard model, clues which next-generation experiments may eventually detect. Two interesting examples are superstrings and non-universal physical laws. We will return to this point in the next chapter.

A more recent but equally poorly understood phase is the one separating the inflation and radiation eras, which is known as thermalization. This is the process which converted the energy that was stored, during inflation, in the vacuum energy density, into the particles that make up the Universe today. One therefore knows that such a process must have occurred if inflation did, but its details (and even the specific physical mechanism behind it) are still poorly understood. Among the particles produced in this phase there were, astroparticle physicists think, those of the dark matter whose presence is indirectly inferred from many astrophysical observations.

Finally, it was only realised in the last 20 years or so that the present Universe is not dominated by matter or radiation. The reason we know this is that around 1998 two groups discovered (around the same time) the Universe is expanding in an accelerated way, which again can't be the case if it is dominated by either radiation or matter: one can show that these always lead to a decelerating Universe. For lack of a better name, the component responsible for this acceleration has been called dark energy, and its observationally inferred gravitational behavior is again very similar to that of a cosmological constant (or vacuum energy density).

This acceleration phase started (in cosmological terms) relatively recently, when the Universe was about one half of its current age, and it therefore seems that the Universe is entering another inflation era. Understanding whether or not this dark energy is indeed a cosmological constant is the most profound problem not only of modern cosmology but of all modern physics, for reasons that we will discuss later in this chapter.

7.2   Key Observational Tests

Among the many successes of the Hot Big Bang model three are crucial, both because they are simple but unavoidable consequences of the model and because historically it was the successive observational confirmation of each of them that eventually led to the model becoming accepted as the standard cosmological paradigm.

The first of these to be confirmed observationally has already been discussed in the previous chapter: it is the Hubble–Lemaître law. To recall the earlier discussion, the Universe is expanding with the galaxies at intermediate distances moving away from each other with a speed proportional to the distance between them. At smaller distances peculiar velocities may dominate (as in the case of the Andromeda galaxy and our own moving towards each other), while at large distances the exact behaviour depends on several properties— including the amount of radiation, matter, and dark energy that it contains. But the law does apply on the scales to which Slipher, Hubble, and Humason had observational access in the early twentieth century.

Notice that this doesn't mean that our galaxy is the centre of the Universe. In fact, there can be no centre of expansion if, as we have assumed, the Universe is homogeneous and isotropic. And this assumption is in very good agreement with a vast range of modern cosmological observations. The easiest way to

$$E = mc^2 + \frac{p^2}{2m} + \cdots , \qquad (5.9)$$

understand this is with the classic analogy of a balloon (in other words, a two-dimensional surface) being filled up. For a creature that lives on the surface of the balloon (such as an ant), there are only two dimensions, and it is impossible for it to see the expansion from the outside.

Nevertheless, an intelligent ant can figure out that it is living in a curved space: if it keeps moving on a straight line it will eventually come back to its starting point (which would not happen in flat Euclidean space). The crucial point to bear in mind is that for the ant it makes no sense to talk about the interior, exterior, or centre of expansion of the balloon. In the case of the expansion of the Universe we need to add one more dimension and think of a three-dimensional surface, but for the same reason it makes no sense to talk about these for the Universe.

It's also important to recall another point that we have already made: the expansion of the Universe is an expansion of space itself, not an expansion of the galaxies in a pre-existing space (which would be a classical, Newtonian view). In some sense, one might say that space is created as the Universe expands. For analogous reasons, it is not correct to imagine the beginning of the Universe as an explosion somewhere in space, for three different reasons that by now you may be able to infer. The first is that there can be no privileged point—remember, the Universe is homogeneous. The second and related one is that the concept of explosion implies a pressure gradient, which again is not allowed. And the third is that space and time did not previously exist—spacetime emerged at that point. So the answer to the question of what existed before the Big Bang is nothing, much in the same way as the answer to the question of what exists north of the North Pole is nothing.

The second observational success of the Hot Big Bang model is the prediction of the primordial abundances of the light elements, also known as primordial nucleosynthesis or sometimes Big Bang nucleosynthesis. This is the result of a series of nuclear reactions which occurred when the Universe was up to three or four minutes old, and the abundances of the relevant elements can be predicted by the model in a rather direct way. To a first approximation (and assuming standard physics) there is only one free parameter affecting the primordial abundances, which is the amount of matter available at that time, in the form of protons and neutrons.

The result of these calculations is that the Universe was initially made of about 75% hydrogen, 24% helium, and 1% of everything else. The fact that this 'everything else' is such a minor contribution often leads classical astronomers to call them simply 'heavy elements', or even more crudely, 'metals'. This terminology can be found both in the cosmological context and when discussing chemical compositions of stars or galaxies—hence the term metallicity. Of these isotopes only deuterium and helium-3 (with about one atom for each ten thousand of hydrogen) and lithium (one atom per ten thousand million of hydrogen) were synthesised at the beginning of the Universe. Anything heavier was produced later on, during the lives and deaths of stars.

Finally, the third observational success is the existence (and main properties) of the cosmic microwave background. The Universe is currently about 13.8 billion years old, and its average density is only about three atoms per cubic meter, but the fact that it has been expanding means that it was much smaller and denser in the past. When it was about 380,000 years old electrons combined with nuclei making the Universe electrically neutral (although this was only a temporary process). Since by definition nothing can escape from the Universe, these photons are still all around us: there are about 410 million of them per cubic meter.

This radiation was discovered experimentally in 1964, by Arno Penzias (b. 1933) and Robert Wilson (b. 1936) using the Holmdel Horn Antenna, which is shown in Fig. 7.3. Interestingly, there were also some detections in the early 1940s, which at the time were not recognised as such. One reason for this is that at that time there was still no known theoretical reason for expecting such a cosmic background radiation to exist. As it turns out, these theoretical predictions appeared just a few years later, but unfortunately nobody connected the two.

Following the discovery in the 1960s the cosmic microwave background has been studied extensively (both from the ground, with high-altitude balloons, and with dedicated satellites) and its properties are now very well understood. For example, it has an essentially perfect black body spectrum (see Fig. 7.4).

As the Universe expands the characteristic temperature of these photons decreases, and at the moment it is about 2.73 °K, or equivalently about −270 °C. This experimental detection of the cosmic microwave background was the third key step in the consolidation of the Big Bang model. The characteristic wavelength of this radiation is around the microwave range. These wavelengths are also close to the range detected by standard (analog, not digital) TV sets, and indeed about 6% of the static noise you see when you try to tune an old (pre-digital) TV set is due to this primordial radiation.

While to a first approximation one measures the same temperature of 2.73 °K in every direction, when looking more carefully one finds slightly different temperatures in each direction. These temperature fluctuations, which have amplitudes of the order of microkelvin (that is, one part in one hundred thousand smaller than the baseline value of 2.73 K) correspond to the primordial density fluctuations that are thought to have been generated

Fig. 7.3 Arno Penzias and Robert Wilson stand at the 15 m Holmdel Horn Antenna, with which they detected the cosmic microwave background. Credit: NASA

during the inflation phase and led to the current large-scale structures in the Universe.

One such map, from ESA's Planck satellite, can be seen in Fig. 7.5. Note that this is actually a remarkable map, being essentially a picture of the whole visible Universe when it was 380,000 years old and became transparent. From that moment the photons that led to this map travelled freely through space without interacting with other particles, until they ended up falling into the focal plane of the Planck satellite telescope.

These observations lead to a wealth of information about the Universe, including its contents (e.g., the amounts of matter, radiation, and dark energy), its geometry (or local curvature), and the properties of the primordial density fluctuations that started the process of structure formation. In fact, thanks to the results of the Planck satellite and other ground and space-based experiments, the cosmic microwave background has been the most sensitive cosmological probe in the last decade. Whether or not this remains the case

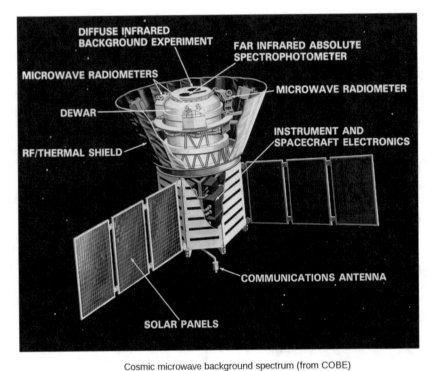

Cosmic microwave background spectrum (from COBE)

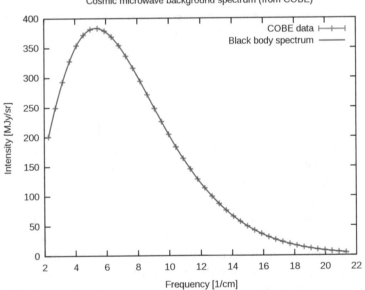

Fig. 7.4 Top: Diagram of the COBE spacecraft, including the FIRAS instrument. Credit: NASA. Bottom: The cosmic microwave background spectrum measured by the FIRAS instrument (1989–1990). The plotted error bars are grossly exaggerated: the real ones would be too small to be seen, and the measured values perfectly match the theoretical curve (Public domain image)

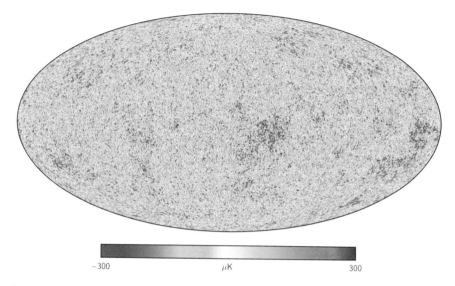

Fig. 7.5 The cosmic microwave background anisotropies seen by ESA's Planck satellite. Credit: ESA and the Planck Collaboration

in the coming years will depend on the outcome of several proposals for next-generation space missions.

Just as a curiosity, you might ask what is the size of the visible Universe. Astronomers often use the distance that light can travel in a given time as a measure of that distance. For example the Moon is about 1.3 light-seconds away, the Sun is 499 light-seconds away, the Andromeda galaxy is 2.5 million light-years away, and so on. Knowing that the Universe is about 13.8 billion years old, you might think that a photon that was emitted right after the beginning of the Universe and which has just reached us now must have traveled a distance of 13.8 billion light-years. However, this is not quite correct, since space has been expanding during the photon's trajectory. Therefore, in order to calculate what distance it has actually traveled one also needs to know how much the Universe has actually expanded in each of the phases of its evolution, and this depends on its detailed contents. Doing the actual calculation, one finds that when the photon reaches us it has traveled a distance that is more than three times greater than in the absence of expansion, specifically, about 46 billion light years.

Another frequent doubt concerns the scales affected (or not) by the expansion of the Universe. For example, are the atoms in your body, or the Earth, or our galaxy, being stretched by this expansion? In all three cases the answer is that they are not, for two separate reasons. The first is that gravitation is the weaker of the four fundamental forces (why it is so much weaker is in fact a very interesting open question), so any objects bound by the other forces are not affected at all by the expansion. This is the case of the atoms in our bodies, which are kept together by electromagnetic interactions. And the second reason is that, even on scales where only gravity plays a significant role, one can have coherent objects (stabilised by their own gravitational fields) or objects interacting with others in their cosmic neighbourhoods through their mutual gravitational fields which are still strong enough to beat expansion. So galaxies are not expanding (on the assumption that they have some dark matter, as we will see shortly), while the best example of the second point is—once again—the fact that the Milky Way and the Andromeda galaxy are approaching each other, rather than moving away, due to their mutual gravitational field. In the present Universe, the scale at which the effects of expansion become noticeable is that of clusters of galaxies.

> The evolution of the world can be compared to a display of fireworks that has just ended: some few red wisps, ashes, and smoke. Standing on a well-chilled cinder, we see the slow fading of the suns, and we try to recall the vanished brilliance of the origin of worlds.
>
> Georges Lemaître (1894–1966)

7.3   The Dark Side of the Universe

Despite its many successes, there are various important questions which the Big Bang model is unable to address. In some sense, they are all related to the choice of initial conditions for the Universe. Here it is important to bear in mind that when the term 'Big Bang' is used colloquially one is often, implicitly or explicitly, thinking of the beginning of the Universe and the (incorrectly) supposed explosion that has started it. But for cosmologists the term 'Big Bang' model refers to the full description of the evolution of the Universe, from the beginning until the present day.

In a pragmatic way, the model can be seen as a computational tool which, given some choice of initial conditions, enables cosmologists to make detailed predictions for the recent behavior of the Universe. These predictions can then be compared to our astrophysical and cosmological observations. However, by itself the model is unable to select the initial conditions that will eventually produce the 'correct' Universe—in other words, the one that we actually observe. In that sense, it is an incomplete model.

The crucial question here is why (and how) were our initial conditions chosen. Perhaps there is some symmetry of Nature (of some conservation law) that we do not yet know which makes this choice the only physically viable one, in the sense that any other choice would be inconsistent with that symmetry. Or perhaps there is a range of possible choices, and the actual choice was made randomly, once and for all throughout the Universe. Or possibly this choice was made many times, independently in various different and sufficiently distant regions of the Universe, such that if you could look sufficiently far away to find one of those different regions you would see a region that not only has different physical properties from the ones in our neighbourhood but even has completely different laws of physics. (In this last case it is to some extent a question of semantics whether one wants to call all those distant regions different universes or still say that they are simply distant—and very different—parts of our own Universe.)

Each of the possibilities mentioned in the previous paragraph is perfectly compatible with all the physics that we currently know, but at the moment we have no way of distinguishing them. While in the third case one could conceivably find supporting evidence (astrophysical tests of the universality of the laws of physics can be done, as we will discuss in the next chapter), for the other two it is not even clear what specific observational fingerprints they may have. The unavoidable conclusion then is that the Hot Big Bang model, as it currently stands, does not fully explain why the Universe has the properties we observe, as opposed to others which (at least in principle) would also be logically possible. It is worth exploring this point further in two other specific cases.

Firstly, let us consider the inflationary phase that we think was responsible for generating the primordial density fluctuations that eventually led to the large-scale structures that we observe today. In principle, one would require models of inflation to satisfy two simple but equally important requirements:

They should be natural and well motivated from the point of view of particle physics, for example, being obtainable from an underlying fundamental paradigm (such as superstring theory).

They should be able to make quantitative predictions that are in at least reasonable agreement with the available astrophysical and cosmological observations.

The problem is that, although there are currently at least hundreds and possibly even thousands of known models of inflation, the list of those which simultaneously fulfill both of the above criteria is the following:

In other words, so far we do not know a single model of inflation that fulfills both of them. Many models of inflation can be derived from fundamental physics, but the predictions made by almost all of them (and especially those that are considered to be the simplest and most natural ones) are in clear disagreement with observations. On the other hand one can use observations, such as data from the cosmic microwave background and the more recent Universe, to build (entirely by hand) simple models that fit the data very well. Such models are often called phenomenological models, but the problem is that one can't easily see how they could stem from fundamental physics, being quite different from the natural ones therein. So although we certainly know, from observations, what inflation is supposed to do, we do not yet specifically know how it actually does it.

But the most obvious example of this issue of initial conditions can be seen in the composition of the Universe, and what is usually called the dark side of the Universe. Only a small fraction of the matter in the Universe emits light and can therefore be seen directly. This has been known for a long time, since pioneering work on galaxy clusters done by Fritz Zwicky (1898–1974, who actually coined the term 'dark matter') in the 1930s and later explored in more detail by Vera Rubin (1928–2016) in the context of what became known as the problem of galaxy rotation curves (see Figure 7.6).

Before discussing this, it is worth starting with a somewhat similar context closer to home. The Earth orbits the Sun with an orbital speed of 30 km/s. If you plot the orbital speeds of the various planets as a function of their distance, d, to the Sun, you will find that the further away a planet is the slower its orbital speed, just like in the previously discussed case of the speeds of satellites orbiting the Earth. Specifically, one finds the proportionality

$$v \propto \frac{1}{\sqrt{d}}, \tag{7.1}$$

which those familiar with secondary school physics can easily derive. This will occur as long as the central object has most of the mass of the system, as is certainly true in the case of the Sun—which has about 99.8% of the mass of the Solar System.

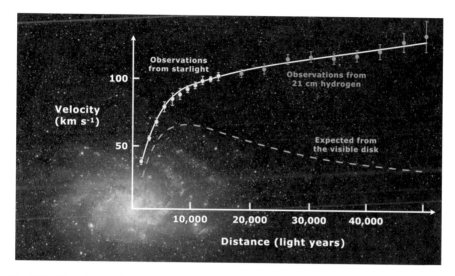

Fig. 7.6 The observed rotation curve of the galaxy Messier 33 (yellow and blue points with error bars), compared to the prediction from distribution of the visible matter (dashed line). Credit: Mario De Leo

In the case of a galaxy one can analogously plot the orbital speeds of stars (or, further away, gas clouds) around the centre of the galaxy. In this case the mass distribution is more spread out, and given the matter that we can seen in spiral galaxies (like our own) one would expect a linearly growing speed close to the centre,

$$v \propto d . \tag{7.2}$$

For very large distances, when it is a reasonable approximation to assume that most of the mass is at the centre, equation (7.1) should again apply. Instead, what one observes is that close to the centre the linear growth does exist but is faster than predicted, while far away the rotation speed is approximately constant rather than decreasing:

$$v \propto const . \tag{7.3}$$

Here one can establish an analogy with the previously discussed issue of the discrepancies in the observed and predicted orbits of the planet Uranus. One possible explanation is that the laws of physics as we know them in the nearby Universe do not apply to these more distant objects. That said, do notice that the rotation speeds are small enough that Newtonian physics should

Fig. 7.7 The Starry Night, painted in 1889 by Vincent van Gogh (Public domain image)

be sufficient to describe them (in other words, relativistic corrections should certainly be very small in this case). The other possibility is that galaxies contain some additional matter which we do not directly see but whose gravitational attraction significantly contributes to the rotation curves.

It is thought that each galaxy that we can see is surrounded my a much larger halo containing this dark matter, whose gravitational effects we detect indirectly—not only in these measurements but in others already discussed, including the cosmic microwave background and gravitational lensing. Although when you see the Andromeda galaxy with your naked eye it is only a small cloud, if you obtain a long-exposure photo of it you will see that it is actually quite large on the sky—in fact larger than the full Moon. This would be even more so if all the matter in the Universe emitted light. In that case, there would be between 100 and 1000 objects in the night sky whose angular size would be larger than that of the full Moon, so the night sky would be very similar to what was imagined by Vincent van Gogh (1853–1890, see Fig. 7.7).

So what is the Universe made of? Everything we can directly see constitutes only about 1% of the total contents of the Universe. Another 4% or so is normal matter (that is, protons, neutrons, and electrons) which we can't see. A lot of this is in the form of intergalactic gas, but some is also thought to be in a form which cosmologists generically call MACHOs (for Massive Compact Halo Objects), meaning comparatively large astrophysical objects such as white or brown dwarfs, neutron stars, or black holes. We are made of this ordinary matter, and to that extent it is of course important for us, but it plays a very minor role in the evolution of the Universe, and in particular in the formation of its large-scale structures.

About 25% of the Universe is in a form known as cold dark matter, and it is thought that it is this that forms most of the halos surrounding the visible parts of galaxies. Here the term 'cold' simply means 'non-relativistic'; this was by contrast to the alternative possibility of hot dark matter, which was a viable scenario but is now observationally excluded, at least as the dominant component. An example of hot dark matter would be neutrinos. Some cosmologists also envisage an intermediate scenario, known as warm dark matter.

What this cold dark matter is, nobody really knows, because so far it has not been directly detected, despite extensive efforts over several decades. What is known is that it can't be normal matter, that is, any particle that we have seen in a laboratory. For a long time the preferred candidate for most cosmologists has been a new type of particle generically called a WIMP (which stands for Weakly Interacting Massive Particle), a specific example of which are hypothetical particles known as neutralinos.

The reason for calling these WIMPs is twofold. Firstly, in addition to interacting gravitationally (as do all particles), any other interactions they have must have a strength similar to that of the weak nuclear force or even weaker—if they interacted any more strongly than that (for example, electromagnetically), they would be easy to detect experimentally, which manifestly they are not, as demonstrated by decades of experiments. The 'massive' part of the name highlights the fact that one expects them to be elementary particles with masses substantially greater than those of the proton and neutron, bearing in mind that they make up about 25% of the Universe, while protons and neutrons make up only about 5%.

Although WIMPs have for a while been the preferred candidate, the fact that experimental searches have so far failed to detect them has led some cosmologists to consider alternative scenarios. One example involves particles called axions, which are actually extremely light particles (much lighter than electrons, for example). Again significant experimental efforts have been made to try to find them, but so far without success.

Finally, about 70% of the energy of the Universe is in a form known as dark energy. Unlike the dark matter, this component does not cluster around galaxies, or if it does, it does so only very slightly. Instead, it forms a uniform density distribution throughout the Universe. Thus, observationally, this is very similar to the energy density of vacuum, or equivalently a cosmological constant. It is this component that is thought to be responsible for the recent acceleration phase of the Universe. The big question is whether this is exactly a cosmological constant or only something close to it (that is, sufficiently close that so far we haven't been able to distinguish the two). Vast ongoing and planned observational efforts are focused on answering this question, because one way or the other the implications are dramatic.

The concept of vacuum energy density arises inevitably in modern particle physics. In fact, particle physicists do not deal with particles but with degrees of freedom called fields, the particles being excited states of these fields. For example, the Higgs particle is an excited state of the Higgs field, a particularly simple type of field known as a scalar field. Thus, particle physicists actually study the so-called quantum field theories, and all the ones we know about necessarily have a non-zero vacuum energy density. In some sense the quantum vacuum is not completely 'empty', but always has some energy associated with it. For example, virtual pairs consisting of a particle and its antiparticle can be created and immediately annihilate. The reason for this is unknown, and is also a deep unsolved problem.

Thus, particle physicists know that their theories predict a non-zero vacuum energy density, and since this corresponds to a cosmological constant they would also predict that the Universe should have one. Until about 1998, when all astrophysical observations were consistent with a vanishing vacuum energy density, particle physicists would cleverly sidestep the problem. Specifically, they would postulate a new (yet unknown) symmetry of Nature which would force the vacuum energy density to be exactly zero. You might think this is pretty close to cheating, but actually it is acceptable. If such a symmetry does exist, it should have other falsifiable consequences that one can look for, either in the laboratory or in the early Universe.

The discovery of the accelerating phase in 1998, and the corresponding dark energy, changed this scenario. This is indeed very similar to a cosmological constant, but when one compares the value inferred from observations to the one predicted (from calculations) in particle physics, one finds that the former is smaller by a factor of

1 000 000 000 000 000 000 000 000 000 000 000 000 000 000 000 000 000 000 000 000 000 000 000 000 000 000 000 000 000 000 000 000 000 000 000 000 000 000

that is, 120 orders of magnitude smaller than the latter. Thus if this is indeed a cosmological constant the argument of the hypothetical new symmetry is moot, and such a small value so far remains unexplained by all existing theories.

This is the reason why cosmologists want to know whether or not the dark energy is a cosmological constant. If it is, there must be something wrong (or at least seriously incomplete) with quantum field theories, since the calculated results for the vacuum energy density are so far away from the observationally determined value. The alternative, preferred by many cosmologists, is that the dark energy is not due to a vacuum energy density. In that case the alleged symmetry can still be invoked to force the true vacuum energy density to be zero, but a different mechanism will be needed to provide the dark energy and the corresponding acceleration of the Universe.

Several such mechanisms exist in principle, and the more likely one involves a dynamical scalar field—that is, one that evolves on cosmological scales as the Universe expands. (The Higgs field is also a scalar field, but it only plays an important role on microphysical scales.) The evolution of this scalar field close to the present day can be sufficiently slow to be very similar (at least to a first approximation and for observations with limited sensitivity) to a cosmological constant. But one consequence of such fields is that they will violate the Einstein equivalence principle—effectively, they introduce a fifth force of Nature, so that gravity can no longer be purely geometrical.

Thus the observational evidence for the acceleration of the Universe shows that our canonical theories of particle physics and cosmology are at least incomplete (or possibly incorrect). One or the other—or possibly even both—need to be extended, so there is new physics out there, waiting to be discovered, which may once more change our view of the Universe.

7.4   Scalar Fields and Extra Dimensions

> If you'd like to know, I can tell you that in your universe you move freely in three dimensions that you call space. You move in a straight line in a fourth, which you call time, and stay rooted to one place in a fifth, which is the first fundamental of probability. After that it gets a bit complicated, and there's all sort of stuff going on in dimensions thirteen to twenty-two that you really wouldn't want to know about. All you really need to know for the moment is that the universe is a lot more complicated than you might think, even if you start from a position of thinking it's pretty damn complicated in the first place. I can easily not say words like 'damn' if it offends you.
>
> Douglas Adams (1952–2001)

Scalar fields are interesting because they are the simplest type of degree of freedom. A scalar field associates a given number to each point of space (or spacetime). For example, when describing the weather conditions, the air temperature is described by a scalar field, while the wind speed needs a vector field: you need to specify a magnitude but also a direction, which are the distinguishing properties of a vector. The temperature and wind speed fields are examples of so-called effective fields, because we know that the temperature and wind speed are macroscopic concepts which approximately describe some of the true microscopic properties of air molecules. But at the microscopic level, modern physics uses various types of fundamental fields.

Remarkably, until 2012 we did not know whether or not there were any fundamental scalar fields in Nature. The question was only answered— affirmatively—with the discovery of the Higgs field (and the associated particle) at the LHC. The Higgs field is behind the mechanism that gives mass to all the other particles (and, at a substantially more technical level, has the additional crucial role of making the theory gauge-invariant). The rest of the standard model of particle physics does not rely on scalar fields: the other particles are described by the so-called Fermi spinors, while the interactions (in other words, the forces) between them are represented by bosonic vector fields.

In this context it is interesting to note that general relativity does not make use of any scalar field; instead, the spacetime geometry (or curvature) is entirely described by a tensor field known as the metric. This is actually remarkable, because almost any consistent gravitational theory one might think of constructing, from Newtonian gravity to the theories with extra dimensions that we will discuss shortly, does include one or more scalar fields.

The fact that there is no scalar field is what defines the class of theories called metric theories of gravity—in other words, it is another way of expressing the Einstein equivalence principle.

Given the now confirmed existence of the Higgs field, the natural follow-up question is whether cosmological scalar fields are also present. If they are, they will certainly have potentially observable astrophysical and cosmological effects. An obvious example are the models for the epochs of inflation and thermalization, almost all of which rely on one (or sometimes several) scalar fields, but many others also exist. In fact, among modern theoretical cosmologists it is almost impossible to find someone who has never invoked a scalar field in their work.

The observational search for evidence of cosmological scalar fields is therefore extremely important, not only from a particle physics point of view but also as a test of Einstein's gravity. We will discuss some possible observational fingerprints of scalar fields in more detail in the next chapter. For the moment we will discuss how scalar fields unavoidably arise in theories that attempt to unify the four fundamental forces of Nature, and in particular in theories with extra dimensions.

The discussion in the previous section clearly shows that one must go beyond the Hot Big Bang model. Adding components to the Universe just to make things come out right might remind you of the old days, in Classical Greece, when extra spheres or epicycles were gradually included. In some sense, adding in the dark matter and dark energy also have similarities to mathematically predicting that Neptune should exist, although they have so far not been found. It is possible that, as in the case of Vulcan (or the Copernican model), what we need in order to make progress is a radically new idea.

To describe earlier epochs in the Universe, when it was extremely hot and dense, we also need to invoke some aspects of quantum physics. To be clear, this is not the same as the often cited goal of unifying general relativity and quantum mechanics—indeed there is as yet no quantum theory of gravity. In our description of the Universe the perfect classical gas mentioned early on in this chapter is replaced by a gas of quantum particles, which was initially homogeneous and isotropic. Despite the increased mathematical complexity of such models, the resulting physical picture is still quite simple.

An important trend in the development of physics is the concept of unification. A good part of the importance of Maxwell's equations is that they mathematically unify electricity and magnetism, while simultaneously providing a single physical description, which we call electromagnetism. Moreover, in the latter part of the twentieth century it became experimentally clear that electromagnetism and the weak nuclear force are two apparently different manifestations of a unified interaction, which we call the electroweak force. One may then ask if this interaction can also be unified with the strong nuclear force, leading to the so-called Grand Unified Theories, or GUTs. And to be even more ambitious, one could envisage a further unification with the gravitational interaction.

One possibly radical idea that has developed in parallel with the notion of unification and, moreover, has been thoroughly studied in the last few decades, are extra spacetime dimensions. Indeed, one can show that in order to try to unify the four fundamental forces of Nature and build a theory of gravity that is consistent with quantum mechanics one needs more than the four spacetime dimensions that we are used to. The theory formerly known as string theory, for example, is best formulated in ten spacetime dimensions. And indeed, one of its characteristics, which is as interesting as it is unavoidable, is the fact that it contains a large number of scalar fields, which play a crucial role in the dynamics of the theory.

Note that the dimensionality of space (or spacetime, as the case may be) plays a role in the number and type of physical solutions as well as mathematical structures. A good example of the latter is that in two space dimensions there is an infinite number of regular convex polygons (the triangle, square, pentagon, hexagon, and so on) while in three space dimensions the number of regular convex polyhedra is finite. These are the five Platonic solids (the tetrahedron, cube, octahedron, dodecahedron, and icosahedron) which Kepler used in his first attempt to model the Solar System.

Several examples of this kind can be found in physical contexts. Starting with Newtonian-type gravitational forces, in hypothetical universes with N spatial dimensions, assuming that there is a gravitational interaction that is otherwise identical to the one we know, the gravitational force will be proportional to

$$F \propto \frac{1}{r^{N-1}}. \tag{7.4}$$

For three spatial dimensions this corresponds to the usual inverse square law. Mathematically one can study the behavior of such forces for any number N, even non-integer numbers. Among the interesting results emerging from this analysis is the fact that if $N < 1$ or $N \geq 4$ there will be no stable orbits. Moreover, objects under the action of forces whose magnitude only depends on distance can only have closed trajectories for two specific cases. One is our own $N = 3$, while the other is $N = 0$ (which of course can also be studied mathematically).

Another example is that a semi-classical analysis of the stability of atoms finds that stable atoms can only exist for $N < 4$, while for $N = 2$ all the energy levels are discrete and the typical distance between the nucleus and the electron in a hydrogen atom (known as the Bohr radius) would be half a centimetre, so atoms would be macroscopic in size. Finally, wave propagation is also affected. For $N = 1$ and $N = 2$, one can show that electromagnetic waves can propagate with any speed up to (and of course including) the speed of light, while for $N = 3$, the only possible speed is the speed of light itself, as expressed in the usual Maxwell equations. In other words, for $N = 1$ and $N = 2$ electromagnetic signals emitted at different times may be simultaneously received, making the transmission of information without distortion much harder.

Even though there are at present no firm ideas about how one can go from these theories with additional spacetime dimensions to our familiar low-energy spacetime cosmology in four dimensions (that is, three spatial dimensions plus time), it is clear that such a process will necessarily involve procedures known as dimensional reduction and compactification. These concepts are mathematically very elaborate, but physically quite simple to understand. Even if the true 'theory of everything' is higher-dimensional, one must still find how it would manifest itself to observers like us who can only directly probe four dimensions.

If extra dimensions do exist, an obvious question is where are they, and why do we only have easy access to the usual four? A simple solution is to make them compact (that is finite and periodic) and very small. For example, if you imagine that you are walking along a tightrope, for you the tightrope is effectively one-dimensional: you can walk forwards and backwards along it, but a sideways step will have unpleasant consequences. On the other hand, for an ant atop the same tightrope, the tightrope is two-dimensional: in addition to backwards and forwards movements, the ant can also easily walk around it. There are several mathematical techniques to compactify dimensions, which resort to scalar fields. In particular, one such field, called the radion, describes the characteristic size of each dimension—which is simply its radius, if the extra dimension happens to be circular (which need not be the case).

Perhaps more surprising is the fact that there can be infinite extra dimensions that are not directly accessible to us. This happens in the context of the so-called brane world models. There is ample evidence for the fact that the three forces of particle physics (the strong and weak nuclear forces and electromagnetism) have a behaviour characteristic of the four spacetime dimensions that we are used to. This is confirmed by a plethora of experiments spanning scales from the proton size up to Solar System scales. For different

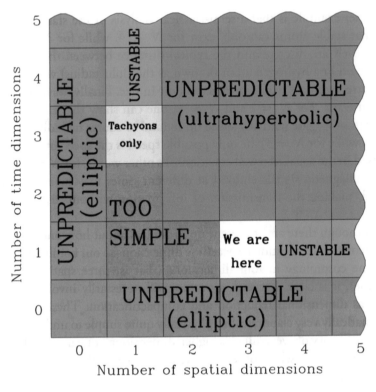

Fig. 7.8 Schematic summary of the properties of n+m-dimensional spacetimes. Credit: Max Tegmark

numbers of dimensions we would either not see some physical processes that have already been seen, or we should have seen some other process that so far has not been seen, as summarised in Fig. 7.8

However, the same is not necessarily true for the gravitational force: Einstein's gravity has only been rigorously tested in local table-top experiments, on Solar System scales, and in specific astrophysical systems such as binary pulsars. In all these cases the gravitational field is fairly weak and for all practical purposes exists in vacuum. Only recently have tests in stronger fields (that is, involving black holes) been started, but for the moment the sensitivity of these tests is limited.

On smaller scales there are only tests of the linear (weak field) regime of gravity, and even so only on scales above one fifth of a millimeter (roughly the thickness of a human hair). This may seem a small scale, but it is many orders of magnitude larger than the size of a proton, or even an atom. The experimental difficulty in testing smaller scales than millimeters is that gravity is so weak that it is very difficult to shield the experiment from electromagnetic and other effects that can be much larger than the ones that are being searched for.

In theories with extra dimensions membrane-type defect objects with several dimensionalities are ubiquitous, and not just with the two dimensions that we commonly associate with the name 'membrane'. In fact, mathematicians generically define n-branes as spatially extended mathematical objects of the corresponding dimensions. (For example, a one-brane is a string and a zero-brane is a point particle.) Thus in typical models with extra dimensions our Universe is a membrane with three spatial dimensions (usually called a three-brane) embedded in a higher-dimensional space that is usually called the bulk. In general, there will be other branes in this space, with various dimensionalities, and again the distance between the various branes corresponds to degrees of freedom described by scalar fields.

The three forces of particle physics are confined to our brane (by a mechanism that is not essential to the present discussion), while gravity and other hypothetical degrees of freedom beyond the standard model (including any scalar fields in the model) are free to propagate everywhere. This would help explain why gravity seems so much weaker than the remaining forces. In these models it is actually not that much weaker, but it only appears to us to be so since we are confined in our three-brane, while gravity is (so to speak) diluted in a much wider volume to which we do not have access.

In this class of models, the 'beginning' of our Universe corresponds to the collision between our brane and a different brane. In particular, it is possible that relevant quantities (such as the density and temperature) would take infinite values at the moment of collision when described by an observer that is confined inside the brane, but they would nevertheless be finite if described by an observer seeing the collision from outside the brane.

What would be the clues for the existence of extra dimensions? One possibility would be the observation of violations of energy conservation in particle accelerator experiments. Collisions at sufficiently high energy can in principle produce the particle mediators of the gravitational interaction (called gravitons, just as the photons are the particle mediators of the electromagnetic interaction), and these could leave our three-brane and escape to the higher dimensional space. If that happened, those gravitons couldn't be detected and neither could the energy they are carrying. Therefore the outcome of the collision would be interpreted as violating energy conservation. The fact that so far no such violations have been detected by the LHC or other particle accelerators puts constraints on the allowed properties of these extra dimensions.

In astrophysics and cosmology there are several contexts in which the presence of extra dimensions may manifest itself. One example are possible changes in the laws of gravitation itself, as expressed by the Einstein equations, whether on sufficiently large or sufficiently small scales (or both). In this type of model, gravitation only has the standard four-dimensional behavior on a limited range of scales, and on higher or lower scales (depending on the specific model in question) the behavior will be different. These differences are related to the fact that in these models, unlike in Einstein's gravity, the metric is not the only field describing the interaction: the scalar fields in the theory also contribute to the gravitational interaction. This is one reason why scalar fields are a possible explanation for dark energy, as mentioned in the previous section.

As pointed out previously, the fact that the metric is the only gravitational field in general relativity and other metric theories of gravity is formally included in the theory through the Einstein equivalence principle. This includes three different components

Universality of free fall (also called the weak equivalence principle): the trajectory of a freely falling test body is independent of its mass, internal structure, and composition.

Local Lorentz invariance: in a local freely falling frame, non-gravitational physics is independent of the frame's velocity.

Local position invariance: in a local freely falling frame, non-gravitational physics is independent of the frame's location.

In a slightly more succinct way one can say that the universality of free fall holds and that the outcome of any local non-gravitational experiment in a freely falling laboratory is independent of the velocity of the laboratory and its location in spacetime.

Additionally, a strong equivalence principle is sometimes used. This contains the Einstein equivalence principle as the special case in which local gravitational forces are ignored. In other words, the strong equivalence principle applies to all bodies (including the self-gravitational interaction of extended bodies) and all physics (including bodies under gravitational forces). While all metric theories satisfy the Einstein equivalence principle, Einstein's general relativity is the only metric theory of gravity that is known to satisfy the strong equivalence principle.

In higher-dimensional theories, new mathematical degrees of freedom and corresponding physical mechanisms naturally appear which necessarily violate the Einstein equivalence principle at some level. This includes scalar fields from string theory, vector fields from some particular extensions of the standard model, and (perhaps even more violently) breaking of Lorentz invariance due to deviations from the spacetime continuum in some suggested quantum gravity models.

Einstein equivalence principle tests are currently the most sensitive probe of new physics, and this will remain the case in the coming years. Each of its three aspects can be separately tested, using specifically optimised experiments. Until recently, these tests—at least the direct ones—were almost exclusively carried out in the laboratory, or exceptionally at low altitude, using balloons or airplanes. More recently, a generation of space tests has started, in which tests of the weak equivalence principle are carried out in LEO and significantly improve on ground-based tests. The advantage of a satellite for these tests is that the free fall times in orbit are much longer than the ones that can be achieved in ground-based experiments.

Specifically, for tests of the weak equivalence principle, the relevant experimental quantity is known as the Eötvös parameter, named after Loránd Eötvös (1848–1919, see Fig. 7.9), who did pioneering work on these tests, using torsion balances, at the beginning of the twentieth century. Remembering that in Einstein's gravity the inertial and gravitational masses are identical, one can quantify possible violations of this principle by defining, for two particular test masses (call them A and B), the difference of the ratios of their gravitational and inertial masses divided by their average, that is

$$
\eta = 2 \frac{\left(\frac{m_g}{m_i}\right)_A - \left(\frac{m_g}{m_i}\right)_B}{\left(\frac{m_g}{m_i}\right)_A + \left(\frac{m_g}{m_i}\right)_B} .
\tag{7.5}
$$

Currently, the most stringent local bound on this parameter, recently obtained in LEO aboard the MICROSCOPE satellite, is

$$
\eta = (-0.1 \pm 1.3) \times 10^{-14} .
\tag{7.6}
$$

Local tests using torsion balances or lunar laser ranging experiments are about one order of magnitude less stringent than this. Either way, all these tests clearly indicate that any violation must necessarily be extremely small, but as we will see in the next chapter astrophysical measurements are even more sensitive.

Fig. 7.9 Loránd Eötvös, photographed in 1912 (Public domain image)

An even more interesting consequence of extra dimensions is the fact that in these models the true fundamental constants of Nature are defined in the higher dimensional spacetime of which our four-dimensional spacetime Universe is a part. What we interpret (and measure) to be fundamental constants are nothing more than effective constants, which are related to the true fundamental constants through one or more scalar fields (depending on the specific model). During the evolution of the Universe these scalar fields evolve in a non-trivial way. It is therefore natural—and indeed inevitable in models like string theory—that there should be small spacetime variations of the constants of Nature that we should be able to measure. If so, this would imply that the laws of physics are in some sense local. We will discuss this in more detail in the next chapter.

What philosophical conclusions should we draw from the abstract style of the superstring theory? We might conclude, as Sir James Jeans concluded long ago, that the Great Architect of the Universe now begins to appear as a Pure Mathematician, and that if we work hard enough at mathematics we shall be able to read his mind. Or we might conclude that our pursuit of abstractions is leading us far away from those parts of the creation which are most interesting from a human point of view. It is too early yet to come to conclusions.

Freeman Dyson (1923–2020)

8

New Frontiers

Here we briefly discuss some open issues and frontiers of current research in cosmology and related areas. The reader is warned that many such issues could be discussed, so the choice of some of them is somewhat biased by my own direct involvement in them. I start by discussing two possible observational manifestations of fundamental scalar fields: the formation of topological defects and the non-universality of physical laws. I also look at the exciting (but not yet fulfilled) possibility of seeing the Universe expand in real time. Finally, in a broader context, I will briefly discuss the peculiarities of the quantum world and the search for life in the Universe.

8.1 Topological Defects and Cosmic Palaeontology

> String theory cosmologists have discovered cosmic strings lurking everywhere in the undergrowth.
>
> Tom Kibble (1932–2016)

In the previous chapter we motivated the search for fundamental scalar fields in the Universe, and also the need to better understand the physics of the earlier stages of the Universe. Topological defects are among the best-motivated consequences of cosmological scalar fields and, if we can detect them, would

© Springer Nature Switzerland AG 2020
C. Martins, *The Universe Today*, Astronomers' Universe,
https://doi.org/10.1007/978-3-030-49632-6_8

provide unique clues about the high-energy Universe. Indeed, they can be thought of as fossils of the earlier stages of the Universe's evolution.

In a certain physical sense, the early Universe is actually much simpler than say the interior of a star. This turns out to be common in physical systems. At low temperatures we tend to have complex structures, but at higher temperatures we have simpler and—importantly—more symmetric ones. It is thought that the Universe was formed in a highly symmetric state, and as it expanded gradually lost some of these symmetries. The concept of symmetry, and more specifically that of spontaneous symmetry breaking, is crucial in modern theoretical physics. Its most important feature is that symmetries are not lost in a slow and continuous way, but in several specific and comparatively short periods, which physicists refer to as phase transitions. It is therefore important to understand how these symmetry breakings occur.

An important concept is that the mathematical theories that describe individual particles or whole physical systems (whether they are classical or quantum mechanical) sometimes possess symmetries that are not shared by some of their solutions. For example, Newton's laws of gravity possess a spherical symmetry (the gravitational force between two objects depends only on the distance between them, and not on their relative orientation in space) and one would therefore expect its solutions to lead to circular orbits. Indeed, these orbits do exist, as do elliptic, parabolic, and hyperbolic ones. It happens that different orbits have different energies, and the lower-energy ones are the elliptic ones, and this is why the orbits of the planets around the Sun have this form. Since it's particular solutions that describe the real world, some underlying symmetries are not seen explicitly, and are said to be broken.

So how does a symmetry get broken? We can understand it with a simple analogy. Imagine a dinner table where napkins are placed between plates. Moreover, suppose the dinner guests have forgotten all social conventions and therefore have no particular preference between the napkins on their left and right. At the beginning of the meal there is a left–right symmetry in the distribution of napkins (physicists would call this a parity symmetry). However, as the meal progresses somebody will want to use a napkin, and will therefore need to choose between them. This choice will affect that of the nearest neighbours, until everybody has made their choice. At this point we say that the symmetry has been broken.

Something similar takes place in a wide range of physical systems, from water (to which we will come in a moment) to the early Universe itself. Here the key variable is not time, but temperature. As temperature decreases, physical systems tend to evolve from symmetry states to states with broken symmetries. The main difference is that in our example we had a discrete symmetry (there were only two possible choices) and in most physical systems the symmetries are continuous: there is an infinite number of possible choices.

However, the above is not the only possible outcome. Imagine that at the centre of our dinner table there is something that prevents people from seeing those sitting across the table. It is quite possible that two people sitting quite far apart make different choices around the same time. If that happens, then somewhere at the table someone will end up with two napkins, while someone else will have none. These regions where there are more or less napkins than one may expect are called topological defects. For physical systems, topological defects essentially correspond to specific energy distributions.

One may argue that it would be possible to arrange things so that everybody agrees to make the same choice. However, if the meal is quite short or if the table is quite big (or even infinite), there won't be enough time to convince everybody to make a similar choice of napkin. Moreover, once someone makes a choice of napkin this becomes irreversible very quickly—nobody will want to swap a used napkin. Again, the same is true for physical systems as the temperature drops, and in the case of the early Universe the speed of light itself sets a limit to the size of regions that can make the same choice, The end result is that topological defects will necessarily form at phase transitions in the early Universe. This process is known as the Kibble mechanism, named after Tom Kibble (1932–2016, see Fig. 8.1).

All you need to produce your own topological defects is a good fridge. Water is a homogeneous liquid (it has no preferred direction), whereas ice crystals are planar, and its formation therefore selects the direction perpendicular to the crystal plane as a preferred one, thus breaking the symmetry. If the freezing process is slow, then all the crystals will line up in the same way, since this configuration is the one with the lowest energy—and Nature always strives for minimum energy configurations. On the other hand, if the freezing is sufficiently fast, in distant regions the crystals will align in different ways, as it won't be possible to agree on a choice of orientation (which is analogous to our choice of napkins at an infinite dinner table). In the borderline zones between regions with differently oriented crystals there will be fracture lines, which are simple examples of topological defects. Producing these defects in the nearest available fridge is left as an exercise for the keen reader.

Topological defects can have a wide range of forms, depending on the types of symmetries involved and on the way they are broken. However, as the previous paragraph suggests they are not exclusive of the early Universe. They form (and have been seen and studied) in a broad range of physical systems, most notably in condensed matter physics. Examples of such systems

Fig. 8.1 Tom Kibble (photographed in 2010), one of the founders of modern particle cosmology, and among many other contributions a pioneer of the study of the cosmological consequences of topological defects (Public domain image)

where they have been studied include liquid crystals, superconductors, and superfluids.

There are three basic types of topological defect:

Domain walls are two-dimensional objects (in other words, two-branes) which form when a phase transition breaks discrete symmetry.

Cosmic strings are one-dimensional or line-like (one-brane) objects which form when a phase transition breaks an axial or cylindrical symmetry.

Monopoles are zero-dimensional point-like (zero-brane) objects which form when a phase transition breaks a spherical symmetry.

There are also various other creatures in the topological defect zoo which are actively studied in cosmology.

Topological defects are highly non-linear objects. While in condensed matter systems they can be studied in the laboratory, in cosmology they have not yet been detected. So their study must rely on a combination of numerical simulations and analytic modeling. As it turns out, the cosmological consequences of each of them are quite different for different types of defects.

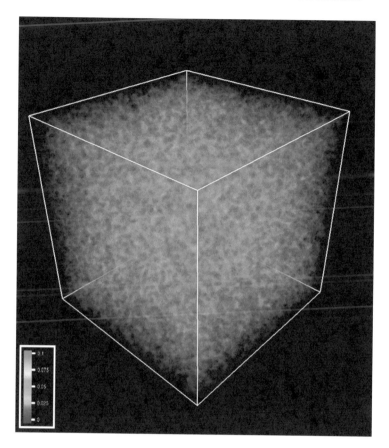

Fig. 8.2 Early evolution of a domain wall simulation box, shortly after the phase transition that produced them. Colors indicate regions of higher or lower energy (Image from the author)

Domain walls are the simplest topological defect: they can be mathematically described by a single scalar field. As such they are also the easiest defects to simulate numerically. Two snapshots of one such simulation are shown in Figs. 8.2 and 8.3.

A network of domain walls effectively divides the Universe into various cells, much like a network of many soap bubbles. (In fact, the physics of domain walls and that of soap bubbles share some interesting similarities.) Domain walls do have some interesting and non-intuitive physical properties. For example, the gravitational field of a domain wall is repulsive rather than attractive.

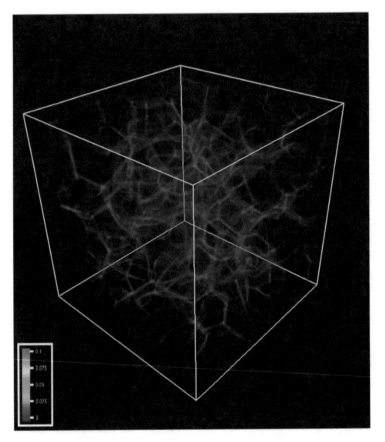

Fig. 8.3 The same simulation box as in Fig. 8.2, shown at a later time. Now the visualisation highlights the boundaries between the wall cells. The intersections of these walls are cosmic strings, which in turn meet at monopole-type junctions (Image from the author)

For this reason, some years ago it was thought that a network of domain walls might be able to explain the recent acceleration of the Universe. However, our own work at CAUP has shown that this is not the case. In a nutshell, it turns out that even in the most favorable circumstances they are only about two thirds as repulsive as a cosmological constant, which current observations would easily distinguish from a cosmological constant. In other words, a domain wall network is not sufficiently repulsive to agree with current observations (which require that dark energy is gravitationally very similar to a cosmological constant). Incidentally, a network of slowly-moving domain walls can indeed be thought of as a cosmological constant confined to two spatial dimensions.

A network of domain walls evolves by gradually increasing the typical size of its domains, and eventually it would dominate the energy density of the Universe (in disagreement with observations, as has already been mentioned). For this reason particle physics models where domain walls form are tightly constrained: any such walls have to be sufficiently light-weight, which means that the phase transitions that produced them can only have occurred recently. The same process of domain growth can be seen in various everyday physical systems. Apart from the already mentioned soap bubbles, another example are tafoni (see Fig. 8.4), cavities which typically evolve by abrasion due to various weathering processes, from wind, water salinity, and temperature changes. They can often be seen on beaches and have also been found on Mars.

Another example of a cosmologically dangerous defect are monopoles. Their existence is actually an inevitable prediction of the Grand Unified Theories (GUTs) which we mentioned in the previous chapter. In other words, given our current knowledge of particle physics we expect them to have formed in at least one phase transition in the early Universe. Moreover, they are predicted to carry an individual magnetic charge, which justifies their name. All magnets we have seen so far are dipoles, having a north and a south pole, and this will still be the case if we break a magnet into two smaller ones.

In fact, GUT monopoles are supermassive, to the extent that having a single one in the visible universe would disagree with current observations. The question of how such monopoles disappeared is one of the puzzles of modern particle cosmology. They may have simply been diluted by the inflation phase (in other words, this phase might have lasted long enough to make the probability of finding one such monopole in the visible Universe extremely small), but it is also possible that they simply annihilated with their antiparticles at some point after their formation.

Finally, the most interesting class of topological defects are the cosmic strings. These can be associated with GUT models, or they can form at the electroweak scale, but one also expects them to form naturally in many models with extra spatial dimensions. A typical GUT string will have a thickness less than a trillion times smaller than the radius of a hydrogen atom, but a 10 km length of one such string will weigh as much as the Earth itself. They can play several useful cosmological roles, including playing a minor part in the formation of structures in the Universe.

String networks typically evolve by gradually losing energy and diluting, in such a way that the typical separation between them grows as the Universe expands. However, they do not usually dominate the Universe, but tend to have an energy density that approaches a constant fraction of the Universe's total energy density. This is known as the scaling solution, and it also means

Fig. 8.4 Tafoni structures in a building on the island of Gozo, Malta (Credit: Dr. Suzanne M. MacLeod) and in Salt Point, California (Credit: Dawn Endico)

that at each epoch one expects to have about a dozen strings passing across the visible Universe (see Fig. 8.5 for two examples) plus some number of small string loops which gradually radiate their energy away and decay.

Various observational efforts are ongoing to detect fingerprints of cosmic string networks astrophysically. So far they have not been found, which again sets constraints on many particle physics models. One of the current

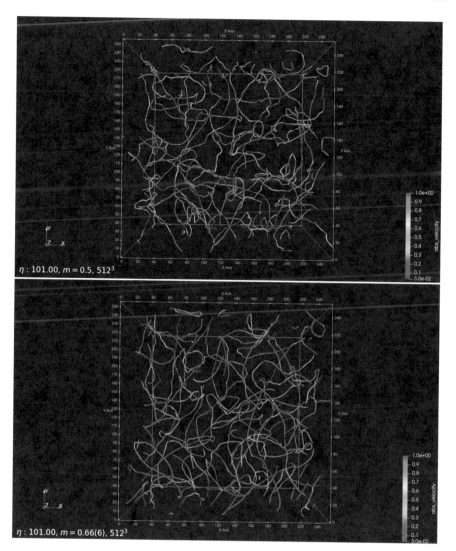

Fig. 8.5 Snapshots of the evolution of a cosmic string network in the radiation and matter eras (top and bottom panels, respectively). The color of the strings denotes their local velocity. Credit: José Ricardo Correia

uncertainties is about the mechanisms by which they actually lose their energy, which affects the detailed properties of these fingerprints, especially in the cosmic microwave and gravitational wave backgrounds. If found, they will provide unique information on the physics of the early Universe.

8.2   Are the Laws of Physics Universal?

One of the most valued guiding principles (or should one say beliefs?) in science is that there ought to be a single, immutable set of laws governing the Universe, and that these laws should remain the same everywhere and at all times. In fact, this is often generalised into a belief of immutability of the Universe itself—a much stronger statement which doesn't follow from the former. A striking common feature of almost all cosmological models throughout history, from ancient Babylonian models, through the model of Aristotle and Ptolemy, is their immutable character. Even today, a small minority of cosmologists still speak in a dangerously mystic tone of the allegedly superior virtues of 'eternal' or 'cyclic' models of the Universe.

In the early stages of the development of physics and astronomy it was thought that different laws of physics applied on the Earth and in the rest of the Universe. Plato and Aristotle, for example, insisted on this dichotomy, which was eliminated in the sixteenth and seventeenth centuries. Interestingly, in many extensions of the standard paradigms of particle physics and cosmology the answer to the question of whether the laws of physics are universal is actually (in a certain sense) negative. This is generically the case in models with dynamical scalar fields and extra dimensions, where the Einstein equivalence principle is violated.

Significantly, this is not only a question left to theoretical speculation, but also one that can be probed experimentally and observationally, and one particularly efficient way of doing it is to test the stability of Nature's fundamental couplings, as already mentioned at the end of the previous chapter. Indeed, these tests are the most powerful way to search for extra dimensions and scalar fields in Nature. One should note that, if string theory is indeed correct, varying fundamental constants and violations of the Einstein equivalence principle are inevitable. Indeed, a detection of varying fundamental constants may well be the only supporting evidence for string theory that can be found in the foreseeable future. Without such experimental evidence, string theory will soon become the modern version of scholasticism.

In Einstein's Universe space and time are not an immutable arena in which the cosmic drama is acted out, but are in fact part of the cast. As physical entities, the properties of space and time—or, more accurately, spacetime—can change as a result of gravitational processes. Interestingly enough, it was soon after the appearance of general relativity and the discovery of the expansion of the Universe—which shattered the notion of the immutability of the Universe—that time-varying fundamental constants were first considered in the context of a complete cosmological model. This was due to Paul Dirac (1902–1984), who apparently worked on this model during his honeymoon, though others had already entertained this possibility.

The so-called fundamental constants of Nature are widely regarded as some kind of distillation of physics. Their units are intimately related to the form and structure of physical laws. Almost all physicists and engineers will have had the experience of momentarily forgetting the exact expression of a certain physical law, but quickly being able to re-derive it simply by resorting to dimensional analysis. Despite their perceived fundamental nature, there is no theory of constants as such, and we don't know what role they play in our physical theories. How do they originate? Are they arbitrary, or fixed by some consistency condition? How do they relate to one another? How many are necessary to describe physics? None of these questions has a rigorous answer at present.

In fact, the only definition of a fundamental constant that we currently have is this:

> A fundamental constant is any quantity whose value can't be calculated in a given physical theory, but must be determined by experiment.

In principle, one may classify different types of constants. Some are applicable only to specific physical objects or processes, for example, the masses of the fundamental particles. Then there are those that apply to classes of physical processes, such as the strength of each of the four fundamental forces. Finally, we have those that in principle are applicable to all physical phenomena, such as the speed of light. Additionally, we may think of special cases such as constants that are (expected to be) zero, such as the mass of the photon, and constants that are (expected to be) unity, such as the ratio of inertial and gravitational masses.

Note that this classification is not rigid. On the one hand, the way a particular constant is classified can change with time, as we develop our understanding of its role. A good example is the speed of light: in Galileo's time and after Ole Roemer's measurement it would have been in the first category in the previous paragraph (a property of specific processes), after the development of Maxwell's electromagnetism it would have been placed in the second category (having been understood to apply to a class of physical processes), while after Einstein's relativity it should be place in the third (being relevant to all physical phenomena). On the other hand, the classification will in general be different in different theories. For example, the ratio of inertial and gravitational masses is only expected to be unity if the Einstein equivalence principle holds.

It is also remarkable to find that different people can have such widely different views on such a fundamental issue of physics. One common view of constants is as asymptotic states. For example, the speed of light c is (in special relativity) the limiting velocity of a massive particle moving in flat (non-expanding) spacetime. Newton's gravitational constant G defines the limiting potential for a mass that doesn't form a black hole in curved spacetime. The Planck constant h is (when divided by 2π) the universal quantum of action and hence defines a minimum uncertainty. Similarly, in higher-dimensional theories such as string theory, there is a fundamental unit of length, the characteristic size of the strings. So for any physical theory we know of, there should be one such constant. This view is acceptable in practice, but unsatisfactory in principle, because it doesn't address the question of the constants' origin.

Another view is that they are simply necessary (or should one say convenient?) inventions: they are not really fundamental but simply ways of relating quantities of different dimensional types. In other words, they are simply conversion constants which make the equations of physics dimensionally homogeneous. This view, first clearly formulated by Eddington, is perhaps at the origin of the tradition of absorbing constants (or 'setting them to unity', as it is often put colloquially) in the equations of physics.

This is particularly common in areas such as relativity, where the algebra can be so heavy that cutting down the number of symbols is a most welcome measure. However, it should be remembered that this procedure cannot be carried arbitrarily far. For example, we can consistently set $G = h = c = 1$, but we cannot set $e = \hbar = c = 1$ (e being the electron charge, and $\hbar = h/2\pi$), since there is then a problem with the fine-structure constant. This constant, also known as the Sommerfeld constant after Arnold Sommerfeld (1868–1951) and defined as

$$\alpha = \frac{e^2}{\hbar c},\tag{8.1}$$

is a dimensionless constant which quantifies the strength of the electromagnetic interaction. The problem is that it would then have the value $\alpha = 1$, whereas in the real world it is measured to be approximately

$$\alpha \sim \frac{1}{137.03599908},$$

(8.2)

with the current experimental uncertainty being only in the last digit.

In any case, one should also keep in mind that the possible choices of particular units are infinite and always arbitrary. For example, at the end of the eighteenth century the metre was defined, thanks to the work of Jean Baptiste Delambre (1749–1822) and Pierre Méchain (1744–1804), as one ten millionth part of half the Paris meridian. Nowadays, it is defined indirectly, such that the speed of light in vacuum is exactly equal to

$$c = 299,792,458 \, \text{m/s}.$$

(8.3)

But this change of definition actually changes the numerical value of the speed of light, as measured in metres per second, by about 0.1%. Obviously, this apparent change has no physical effects. The only thing that has changed is our choice of the measuring ruler and clock that we use.

Perhaps the key point is that there are units which are arbitrary and units which are more relevant at least in the sense that, when a quantity becomes of order unity in the latter units, dramatic new phenomena emerge. For example, if there was no fundamental length, the properties of physical systems would be invariant under an overall rescaling of their size, so atoms would not have a characteristic size, and we wouldn't even be able to agree on which unit to use as the 'meter'. With a fundamental quantum unit of length, we can meaningfully talk about short or long distances. Naturally, we will do this by comparison to this fundamental length. In other words, 'fundamental' constants are fundamental only to the extent that they provide us with a way of transforming any quantity (in whatever units we have chosen to measure it) into a pure number whose physical meaning is immediately clear and unambiguous.

Still, how many really 'fundamental' constants are there? Note that some so-called fundamental units are clearly redundant: a good example is temperature, which is simply the average energy of a system. In our everyday experience, it turns out that we need three and only three: a length, a time, and an energy (or mass). In other words, we need to specify our ruler, our clock, and our weighing scale. However, it is possible that in higher-dimensional theories (such as string theory), only two of these may be sufficient. And maybe, if and when the 'theory of everything' is discovered, we will find that even less than two are required.

A final metrological point that must be stressed is this:

> One can only measure dimensionless combinations of dimensionful quantities, and any such measurements are necessarily local.

For example, if I tell you that I am 1.76 metres tall, what I am really telling you is that the last time I took the ratio of my height to some other height which I arbitrarily chose to call 'one metre', that ratio came out to be 1.76. There is nothing deep about this number, since I can equally tell you my height in feet and inches. Now, if tomorrow I decide to repeat the above experiment and find a ratio of 1.95, that could be either because I've grown a bit in the meantime, or because I've unknowingly used a smaller 'metre', or due to any combination of the two possibilities. And the key point is that, even though one of these options might be quite more plausible than the others, any of them is a perfectly valid description of what's going on with these two measurements: there is no experimental way of confirming one and disproving the others.

Similarly, as regards the point on locality, the statement that 'the speed of light here is the same as the speed of light in the Andromeda galaxy' is either a definition or it's completely meaningless, since there is no experimental way of verifying it. In other words, when making such a statement one must be implicitly defining units of measurement such that this is the case. These points are crucial and should be clearly understood.

In other words, although it is not only possible but even convenient to build models in which dimensionful quantities (such as the speed of light or the electron charge) are variable, there is nothing fundamental in that choice, since any such theory can always be transformed into an absolutely equivalent theory in which a different quantity (rather than the original one) is variable. If that transformation involves an appropriate choice of units the observational consequences of the two theories will be identical.

On the other hand, from a purely experimental (or observational) point of view it makes no sense to measure dimensionful constants per se: one can only measure dimensionless quantities, and perform tests based on those quantities. Two examples of dimensionless quantities are the fine-structure constant which we have already introduced, and the ratio of the proton and electron masses

$$\mu = \frac{m_p}{m_e} \sim 1836.1526734\,, \qquad (8.4)$$

where the experimental uncertainty is again in the last decimal place shown. This ratio is related to the strength of the strong nuclear force.

Searching for experimental or observational clues for varying fundamental constants is a challenging task for several reasons. In addition, to the fact that one can only measure dimensionless quantities, any such variations are expected to be very small, for the obvious reason that if they were significant they would have been easily detected long ago. On the other hand, if they do exist at some level, then almost all of the physics that we currently know is only (at best) approximately correct, and effects of these variations should be ubiquitous.

To give a specific example, if the fine-structure constant α varies this means that the strength of the electromagnetic interaction will vary correspondingly. An immediate consequence will be that all the atomic energy levels and binding energies will be affected. By how much this happens, for a given change in α, will depend on the atom and energy level in question. Similarly, a varying proton-to-electron mass ratio will affect rotational and vibrational energy levels in molecules.

More generally, we can describe qualitatively how some of the basic physics that we are used to would be changed if the strength of each of Nature's four fundamental forces were changed:

If the strength of electromagnetism increases (in other words, if α increases), atoms will become unable to share electrons, so there will be no chemistry; if it decreases, atoms won't be able to retain electrons, and we will only have free particles.

If the strength of the strong nuclear force increases, stars will be able to convert all light elements to iron, leading to a Universe without hydrogen; if it decreases, no complex nuclei will form—and in particular, no carbon will form.

If the strength of the weak nuclear force increases, atomic nuclei will decay before heavy elements can form; if it decreases, all hydrogen will be converted into helium.

If the strength of the gravitational force increases, the Universe will quickly stop expanding and recollapse; if it decreases, the expansion will be so fast that no cosmic structures will be able to form (so the Universe will only contain diffuse particles).

You might think that very substantial changes in the strength of these interactions would be needed in order to see significant effects, but this is in fact not the case—even small changes can have dramatic physical effects. For example, if the strength of electromagnetism were increased by 4% or the strength of the strong nuclear force reduced by 0.4%, the amount of carbon produced in stellar cores would be drastically reduced.

Conversely, if the strength of electromagnetism were reduced by 4% or the strength of the strong nuclear force increased by 0.4%, then the production of oxygen in stars would be greatly reduced. Indeed, there is only a narrow range of the strengths of the two interactions which allows significant quantities of carbon and oxygen (both of which are crucial for life as we know it) to be simultaneously produced.

Further increasing the strong nuclear force, by 4%, the diproton (an isotope of helium whose nucleus contains two protons but no neutrons), would be stable. This would dramatically speed up some nuclear fusion processes and stellar lifetimes would be correspondingly reduced. On the other hand, deuterium would not be produced and therefore carbon and oxygen would not be produced either, as in the previous paragraph. Deuterium also becomes unstable if one decreases the strength of the strong nuclear force by 10%, in which case the periodic table would have only one element—hydrogen. In such a Universe stars would still exist, but the Chandrashekhar mass would increase to about 5.5 solar masses (heavier stars would become black holes).

As for the proton-to-electron mass ratio, μ, if it were made much smaller, then it would become impossible to form ordered molecular structures. Finally, the fact that the neutron mass exceeds that of the proton by only 0.14% is also worthy of note. If the difference increased to 2%, then again no carbon or oxygen would form, while if the neutron mass were smaller than the proton, then hydrogen itself would be unstable.

It is therefore important to identify the physical processes in which a given variation has a larger effect, or at least one that might be more easily detectable given currently available technologies. Usually, the way to proceed is to identify two related physical processes whose dependence on the constant one wants to measure is different. By measuring observables related to the two processes, we can extract a measurement of the constant that is consistent with the observations. We can further discuss specific examples of methods that are used locally (in laboratory experiments) and also in astrophysics and cosmology.

Laboratory measurements of the fine-structure constant have been done for decades, although the currently preferred method (which uses atomic clocks) is a lot more recent. Any clock is composed of two basic parts: something that ticks periodically and something that counts the number of ticks and converts it into a reference unit like seconds. In the overwhelming majority of everyday clocks, the period oscillator is mechanical (as in a pendulum) or electromagnetic (as in a quartz crystal). An atomic clock is distinctive because the oscillator is not classical but quantum mechanical: a photon absorbed by the last electron of an atom, thus changing its spin and magnetic field orientation. Recent developments include laser-cooled atomic clocks, clocks based on a single atom, THz-frequency clocks, and placing these clocks in microgravity (at the International Space Station or in dedicated satellites). So far these measurements have only found upper bounds for the variation of the fine-structure constant, at the level of 10^{-17} per year. More recently, analogous measurements have also started to be made for the proton-to-electron mass ratio with similar results (though they are still weaker by about two orders of magnitude).

As in the case of the weak equivalence principle tests discussed at the end of the previous chapter, a possible improvement in the coming years may come from doing analogous experiments under microgravity. The sensitivity of such tests for possible variations can be several orders of magnitude better than is currently achievable in terrestrial laboratories. This is essentially due to the fact that the precision of this type of clock is inversely proportional to the strength of the gravitational field in which the clock is found. This should offset the obvious practical difficulties related with the placing of clocks in orbit (particularly having to do with calibration).

Quasars are the brightest constant light sources in the Universe: each of them is more than one hundred times brighter than all the stars in our galaxy taken together. It is suspected that the reason for this is that the centre of each quasar has a supermassive black hole. Because they are so bright, it is possible to observe them at extremely great distances. In particular, with modern telescopes we can obtain high-resolution spectra of these quasars. These spectra will contain a set of emission and absorption lines, which are fingerprints for the various atoms and molecules in the quasar itself or in the interstellar medium: emission lines are produced by material in the quasar itself, while absorption lines are produced as the light coming from the quasar on the way to the telescope crosses low-density clouds of gas and dust (for example, at the edge of galaxies) that are so thin that they usually can't be observed directly.

Spectral lines corresponding to each atom or molecule are characterised by wavelengths which are well known, having been studied in the laboratory. The expansion of the Universe while light travels from the quasars to us stretches the wavelengths to longer, redder wavelengths. This process is the previously discussed redshift, and can easily be accounted for in the data analysis procedure. Since the atomic energy levels depend, among other things, on the strength of electromagnetism, if the fine-structure constant had a different value in the past then these energy levels (and the wavelengths of the corresponding lines) will also be changed. In this case they may be redshifted or blueshifted, depending on the transition in question.

Now, if one can measure only one transition it's impossible to separate the effect of the ordinary redshift due to the expansion of the Universe (which effectively provides the distance to the source) and the effect of a possible variation of α. On the other hand, if one can measure two or more absorption lines that have different sensitivities to α and ensure that the two transitions were produced at the same distance (in other words, in the same absorption cloud), then one can distinguish the two effects and simultaneously measure the redshift of the cloud and the value of α at that redshift. Therefore, it is in principle possible to use quasar spectra to measure the fine-structure constant (as well as other constants) at the epoch in which these lines were formed—typically when the Universe had 20–60% of its current age.

Several groups have been trying to carry out this kind of measurement, which require high-resolution data that can only be obtained from the world's largest currently available telescopes, such as the VLT. Some of this data seems to indicate that there is indeed a spacetime variation of the fine-structure constant, at the level of a few parts per million of relative variation. At the moment these results are far from definitive, as can be seen in the top panel of Fig. 8.6, which plots the relative variation of α, viz.,

$$\frac{\Delta\alpha}{\alpha}(z) = \frac{\alpha(z) - \alpha_0}{\alpha_0} \,, \tag{8.5}$$

where α_0 denotes the local laboratory value. More recent detailed observations have both improved the sensitivity of measurements at relatively low redshifts and enabled measurements at significantly higher redshifts, as illustrated in the bottom panel of Fig. 8.6, but the results are still inconclusive.

One bottleneck of currently published measurements is the limited sensitivity of existing spectrographs. A new generation of high-resolution spectrographs (and ideally also telescopes) is needed to settle the issue—as well as independent methods that can corroborate these results. An example is the ESPRESSO spectrograph, that recently became operational at the VLT (see Fig. 8.7), and is for the first time able to combine light from the four VLT unit telescopes, corresponding to an effective collecting area of a sixteen meter telescope. In the coming years, the VLT's main successor, ESO's Extremely Large Telescope (ELT), will enable even more sensitive tests.

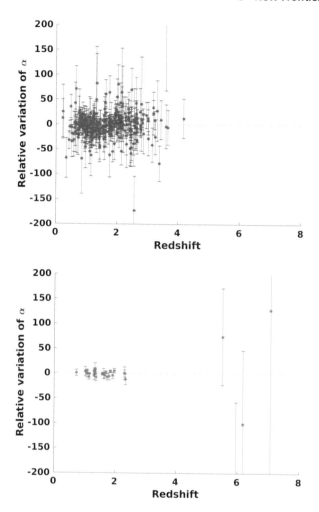

Fig. 8.6 Current astrophysical measurements of the fine-structure constant α, as a function of the redshift. The top panel shows an early dataset by Webb et al., while the bottom panel shows more recent measurements listed in Wilczynska et al., illustrating both the improvements in sensitivity at low redshifts and the extension of the redshift range in which the measurements can be made. The measurements are plotted as a relative variation, as compared to the laboratory value, in parts per million

Observational evidence for the variation of the fine-structure constant and/or the proton-to-electron mass ratio will immediately imply the violation of the Einstein equivalence principle, and the end of the concept of gravity as a geometric phenomenon. Clearly, it violates local position invariance, but a more detailed analysis shows that it also violates the weak equivalence principle (refer to the previous chapter for the definitions of both of these). It would also

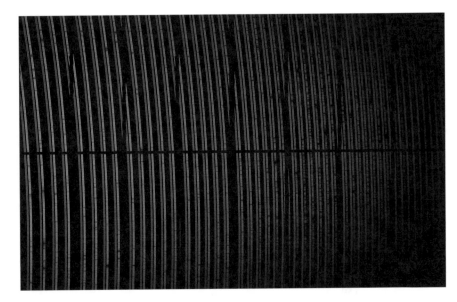

Fig. 8.7 Spectral data from the First Light of the ESPRESSO instrument at ESO's VLT in Chile. The light from a star has been dispersed into its component colours, together with light from a calibration light source. Close inspection shows many dark spectral lines in the stellar spectra. The dark gaps are features of how the data is taken, and are not real. Credit: ESO/ESPRESSO team

reveal the existence of further gravitating fields in the Universe, and provide very strong supporting evidence for the existence of extra dimensions. This would be a more dramatic revolution than the one that replaced Newton's gravity by Einstein's.

Since a varying α would violate the weak equivalence principle, one can use α measurements to constrain the Eötvös parameter defined in the previous chapter. My recent analysis (done together with Maria Prat Colomer), putting together all existing measurements of α along the line of sight of quasars as well as the local ones with atomic clocks and combining these with the direct bound from the MICROSCOPE experiment, leads to the upper bound

$$|\eta| < 4 \times 10^{-15}, \tag{8.6}$$

which is about three times stronger than the MICROSCOPE bound on its own and about thirty times stronger than local direct tests using torsion balances or lunar laser ranging experiments. This shows that astrophysical measurements can seriously compete with local ones.

It is also worthy of note that if variations are observationally confirmed, the comparison between the relative variation of α and that of μ will provide a crucial test of fundamental physics. In particle physics models where the variation of the fine-structure constant is due to the dynamics of a scalar field, the proton-to-electron mass ratio will unavoidably vary, too. In particular, this is the case in GUTs, although the specific relation between the two variations will be different for different models of this class, but this is a blessing rather than a curse. If one can observationally measure both quantities at several epochs in the early Universe, one will be able to distinguish among models of unification. This is a more powerful test of fundamental physics than anything we can hope to do in particle accelerators.

8.3 The Dawn of Real-Time Cosmology

> [...] all models are approximations. Essentially, all models are wrong, but some are useful. However, the approximate nature of the model must always be borne in mind.
>
> George Box (1919–2013)

These high-resolution spectroscopy measurements of the fine-structure constant along the line of sight of bright quasars are also important for cosmology, and in particular for addressing the enigma of the acceleration of the Universe and dark energy, for two different reasons.

The first has already been mentioned. They are an efficient way to ascertain whether the dark energy is due to a cosmological constant or to the dynamics of a cosmological scalar field. In fact, modern cosmologists understand that in any realistic model where dark energy is due to a cosmological scalar field, that field will interact with (or, in technical terms, will be coupled to) electromagnetism, and therefore will also lead to a spacetime variation of α. Thus measurements of α can also be used (together with the traditional cosmological datasets) to constrain dark energy models.

The second and broader point stems from a point that we made in the previous chapter. The comparison between theoretical models and cosmological observations relies on some underlying assumptions, such as large-scale homogeneity and isotropy, or the validity of general relativity on those cosmological scales. Therefore, the results of these comparisons, such as the fact that the present Universe seems to be made of about 30% matter (most of which is dark matter) and 70% dark energy, are, to that extent, model-dependent. Moreover, the physical properties of this dark energy (gravitationally similar to a vacuum energy density with an unexpected value) are particularly odd, so direct ways of probing are particularly desirable.

Also, the fact that most of the contents of the Universe are two dark components, which have been mathematically predicted but not directly observed, may lead one to ask whether they are epicycles that we have temporarily added, destined to be replaced at some future stage. The logical conclusion is that model-independent ways of confirming the acceleration of the Universe, and—more broadly—mapping the expansion history of the Universe are a crucial missing piece to consolidate the current cosmological model or point out how it should be modified and improved.

The redshifts of astrophysical objects following the overall cosmological expansion (in other words, having negligible peculiar velocities), gradually change as a function of time due to the evolution of the expansion rate of the Universe. This effect is known as the redshift drift, or the Sandage test, named after Allan Sandage (1926–2010) who first pointed out the relevance of this effect in 1962.

Detecting this effect would constitute a direct and model-independent determination of the expansion history of the Universe, without any underlying assumptions. It would also provide independent evidence for the acceleration of the Universe and (if you will forgive the phrase) shed light on the nature of the dark energy behind it. Just as the redshift provides evidence for the expansion of the Universe, the redshift drift can provide evidence for the acceleration or deceleration of the Universe: broadly speaking the redshift drift will be positive in the former case and negative in the latter. The rate of drift of the redshift is given by the simple expression

$$\frac{\Delta z}{\Delta t} = H_0(1+z) - H(z), \tag{8.7}$$

where H_0 is the Hubble constant (that is, the present-day value of the expansion rate), while $H(z)$ is the expansion rate of the Universe at the generic redshift z.

The ELT can directly detect and characterise this redshift drift, in a range of redshifts which probe the matter era of the cosmological expansion and would otherwise be difficult to probe observationally. Other astrophysical facilities may also be able to do it at lower redshifts. And it turns out that the bright quasars that provide measurements of the fine-structure constant are also, quite often, ideal targets for the ELT's redshift drift measurements. In practice, what

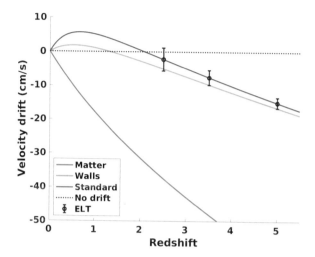

Fig. 8.8 The redshift drift signal, in cm/s, as a function of the redshift of astrophysical objects following the cosmological expansion, over a period of years. The blue curve is for the standard cosmological model, while the red and green curves illustrate alternative models: one containing only matter and another one with domain walls. The three points with error bars illustrate the expected sensitivity of the ELT measurements

the ELT will measure is a spectroscopic velocity, which can be easily related to the redshift drift:

$$\frac{\Delta v}{\Delta t} = \frac{c}{1+z}\frac{\Delta z}{\Delta t}. \tag{8.8}$$

Figure 8.8 illustrates the redshift drift signal as a function of the redshift of astrophysical objects following the cosmological expansion, and the expected sensitivity of ELT measurements. The currently preferred standard model, containing 30% matter and 70% dark energy (assumed to be a cosmological constant) is depicted by the blue curve, showing a positive redshift drift at low redshifts (when the Universe is accelerating) and a negative drift at high redshifts (when the Universe is in the matter era).

On the other hand, the red and green curves illustrate alternative models. The former is for a Universe containing only matter (which never accelerates and therefore always has a negative redshift drift), while the latter is for a Universe where the dark energy is due to domain walls which, as discussed earlier in this chapter, does not accelerate enough.

With the ELT probing the matter era and other astrophysical facilities making similar measurements at lower redshifts, we have the prospect of seeing the Universe expand in real time, obtaining a detailed mapping of its expansion history, and therefore providing a precision consistency test of the standard cosmological paradigm. Whether or not it will pass the test remains to be seen.

8.4   The Quantum World

While I am describing to you how Nature works, you won't understand why Nature works that way. But you see, nobody understands that.

Richard Feynman (1918–1988)

How do you know, in your everyday life experience, that you don't live in a Universe governed by the laws of classical physics? You can build a Wilson cloud chamber as we have already discussed, but there is actually a much simpler way. All you have to do is stick a magnet to the door of your refrigerator. This is the result of the Bohr–van Leeuwen theorem, which Niels Bohr (1885–1962) and Hendrika van Leeuwen (1887–1974) independently discussed in their doctoral theses in 1911 and 1919, respectively. In a slightly simplified way, it states that if you assume nothing more than the laws of classical physics and statistical mechanics, and use them to model a material as a system of electric charges, then the system can have no net magnetization (in other words, it will not be magnetic). So there can be no lodestones in a purely classical Universe.

We have already seen, when discussing relativity, that the way the Universe seems to work is not always in agreement with our classical intuition. Some physical processes have no classical equivalent, and trying to describe them through a classical mental image will therefore be inadequate and may even lead to paradoxes and inconsistencies. We also mentioned—especially when considering the development of astronomy from Copernicus to Galileo—that there is a conceptual difference between a mathematical description of a set of physical processes and a physical model describing these processes.

Both of these are even more explicitly manifest when describing the quantum world. Quantum theory was born in the year 1900, with the pioneering work of Max Planck (1858–1947, see Fig. 8.9), and became consolidated in the 1920s. It is arguably the most successful theory that has been developed so far (some of its predictions have been experimentally confirmed to more than ten decimal places), and underlies a significant part of modern technology. Mathematically, it is comparatively simple: almost everyone that learns both

Fig. 8.9 Max Planck in 1933 (Public domain image)

quantum physics and relativity finds the mathematical techniques relevant for the former easier to learn. Nevertheless, when one tries to build a mental image of what this formalism means (based, of course, on our everyday classical intuition), one fails entirely.

One major difference is that in the quantum world the act of measurement necessarily affects the object being measured. No matter how one does it, there is always a minimum perturbation that a measurement will cause. This is one of the aspects of Heisenberg's uncertainty principle, named after Werner Heisenberg (1901–1976), and it is not a practical limitation from the measurement tools we use—it is a fundamental and unavoidable limitation. For example, you can't measure both the position and velocity of an electron with infinite accuracy. Strictly speaking, this effect also exists for macroscopic objects, but is entirely negligible in that case.

That being the case, are we allowed to assign values (or even to speak) about physical properties whose values we have not determined by measurement? For example, given that we can't measure both the position and velocity of an electron with arbitrary precision, can we say that the electron does have a position and a velocity? Classically, we can certainly imagine it as such, but are we actually allowed to do that?

One may in principle think of various possible answers to the question. One might insist that an electron does have a position and a velocity but as it happens we are simply unable to measure both of them simultaneously, or one might say that there is no reason to assume that the position and velocity, which are certainly useful concepts for macroscopic objects, are applicable to electrons, or indeed to anything in the microphysical world.

Einstein—himself one of the founders of quantum physics—was not satisfied with the latter answer, instead preferring the first. Moreover, he pushed the discussion further by arguing that if quantum theory can't describe both the position and velocity of an electron, then, no matter how good a mathematical description it may be, it is not a complete physical theory since it doesn't provide a complete representation of what an electron actually is.

Einstein's argument was made explicit in what became known as the Einstein–Podolsky–Rosen (or EPR) paradox, from the names of the authors of an article published in 1935. The argument was based on one of Einstein's famous conceptual experiments, which argued that in order to avoid what he called a 'spooky action at a distance' (that is, the propagation of information faster than the speed of light) there ought to be 'hidden variables', that is, properties of physical objects that are not part of the quantum theoretical description. Thus, to that extent the quantum description of Nature is incomplete.

Being a conceptual experiment, there was initially no intention of actually doing it. However, subsequent work by David Bohm (1917–1992) and John Bell (1928–1990) showed that if Einstein's hidden variables hypothesis was correct there was an unavoidable prediction, known as Bell's inequality, which not only was different from the quantum mechanical prediction, but could actually be tested experimentally. Two decades of efforts to carry out the test culminated in the first unambiguous experimental result, provided in 1982 by Alain Aspect (b 1947, see Fig. 8.10) and his collaborators: the predictions of quantum theory were confirmed, and Bell's inequality violated.

What does this mean? There are at least two conclusions to be drawn. The first is that quantum mechanics passed one further experimental test, and although it is conceivable that it will one day be superseded by a different theory whose contents we do not yet know, what we do know is that any such theory cannot be based on hidden variables. Any such theory is experimentally ruled out by the violation of Bell's inequality.

Fig. 8.10 Alain Aspect in 2010 (Public domain image)

The second is what is sometimes called the 'shut up and calculate' approach. Although quantum physics does not offer a clear physical description of the Universe on microscopic scales, it does provide a powerful and accurate computational algorithm to predict the result of any experiments one may want to do. At least to a certain extent, that may be sufficient.

A separate and more profound question is what constitutes an explanation in physics (or, more generally, in science). You will remember that provided deduction of Kepler's empirical laws, as a consequence of his inverse square law. In this case one could argue that he provided a physical explanation for them, but the argument could also be made that he simply provided an algorithm for calculating these orbits. A deeper explanation for this behaviour was subsequently provided by Einstein's gravity.

In science it's always important to know which questions to ask. Profound scientific advances often occur from the realisation that questions formerly considered fundamental were irrelevant and/or should not be asked at all. Again Kepler provides a good example: in his day, the question of why there were six planets in the Solar System seemed fundamental, so it seemed important to provide an answer (and he thought he had found one), but today we understand that this is an irrelevant question, because the number of planets—which of course is no longer six—is due to initial conditions and accidents in the process of formation of the Solar System. Perhaps the deep question here is: what is it in the way we currently think about the Universe that causes us to find the quantum mechanical behaviour (or the value of the cosmological constant, and so on) so puzzling?

> You can't blame most physicists for following this 'shut up and calculate' ethos because it has led to tremendous developments in nuclear physics, atomic physics, solid-state physics and particle physics.
>
> Jean Bricmont (b. 1952)

8.5   The Search for Life in the Universe

> It [the beauty of physics] has nothing to do with human beings. Somewhere, in some other planet orbiting some very distant star, maybe another galaxy, there could well be entities that are at least as intelligent as we are, and are interested in science. It is not impossible. I think there probably are lots. Very likely none is close enough to interact with us, but they could be out there very easily. And suppose they have [...] very different sensory apparatus and so on, they have seven tentacles, and they have fourteen little funny-looking compound eyes, and a brain shaped like a pretzel, would they really have different laws? Lots of people would believe that, and I think it is utter baloney. [...] They [the three principles in Nature] are emergent properties. [...] Life can emerge from physics and chemistry plus a lot of accidents. The human mind can arise from neurobiology and a lot of accidents.
>
> Murray Gell-Mann (1929–2019)

As we explore the Universe, we are also compelled to think about our place within it. For example, the Earthrise photo (see Fig. 8.11) gives us some perspective on our location, a 'pale blue dot' in Carl Sagan's words, and also prompts the question of whether or not we are alone in the Universe.

Today we don't know the answer to that question although, interestingly, there were times in the past when we thought we knew. In the early twentieth

Fig. 8.11 The Earthrise photo, taken by Apollo 8 astronaut William Anders on 24 December 1968. Credit: NASA/William Anders

century, for example, it was taken for granted (to the extent of being covered in full-page news items in mainstream newspapers) that Mars was inhabited by a society with a sufficiently advanced technology to be able to build immense canals to transport water from the polar to equatorial regions. Today, we know that these supposed canals were simply geological structures on the surface of the planet, over-interpreted due to the limited resolution of the telescopes then available and a significant amount of wishful thinking.

Today, the search for life elsewhere is most commonly framed in the context of exoplanet studies. The goal of this search is often described as the search for a 'twin Earth': a planet like the Earth orbiting a star like the Sun. This has the advantage of being easy to explain and understand, but from a physics point of view it is hopelessly narrow-minded, since it is clearly not the best way to answer the question of how common life is in the Universe. This is sometimes called the McDonald's effect: if you're a tourist visiting a new city for the first time, when dinner time arrives do you really want to go to a McDonald's? It's likely that you will find one, but it is not the best way to learn about the place you are visiting. In exoplanet searches as in many other areas, one searches where the light is good (in more ways than one).

One problem we often face when addressing this question is our tendency to be parochial, and assume that something that applies specifically to us is necessarily representative in a wider context. In fact we (humans) are not at all representative of the Earth's biomass: we only make about 0.01% of it (most of the Earth's biomass is actually in plants). Another example of a parochial concept is the notion of extremophiles, beings which live, on the Earth, in conditions of temperature and pressure that are quite different from our own. This example also shows that the concept of a habitable zone is equally parochial.

What, then, is life? Obviously, this is a challenging question, and one difficulty stems from the fact that as far as we known there is no single characteristic property that is simultaneously intrinsic and unique to living systems. Life as we know it uses atoms that are common in our environment and chemical processes that are not intrinsically different from inorganic reactions. A general definition might be something like this:

> Life is a self-sustaining physical system able to transform environmental resources in its own constituents and to evolve by natural selection.

You may be surprised to find that some things are not part of the definition. Perhaps the most obvious is self-reproduction, but there is a good reason for its absence: you can't actually self-reproduce—you need someone else's help. This obvious example shows that self-reproduction is usually necessary at some level (say, at the level of substructures and species, though not individuals), but it is not sufficient.

Importantly, the system must contain information about itself—a sort of instruction manual for how to build a copy—which is preserved by natural selection. But the probability of an exact copy decreases with complexity, which also allows for different rates of evolution.

If we want to be more specific and consider intelligent life, a possible definition might be this:

> Intelligent life is any living system capable of passing a Turing test whose questions include Nature's fundamental laws and structure.

This exploits the concept of Turing test, developed in 1950 by Alan Turing (1912–1954) with the goal of deciding whether a machine could be said to exhibit intelligent behaviour. The idea in our case would be that, if a human being, after spending a sufficient amount of time talking remotely (by exchanging e-mails or text messages, for example) with another human and an alien, can't reliably tell which is which, then one can legitimately say that the alien is intelligent.

From a physics point of view, life clearly requires a thermodynamic imbalance mechanism. In our case this relies on chemical bonds and (ultimately) light from the Sun, but several other energy sources can be envisaged:

Thermal, ionic, or osmotic gradients

Kinetic energy of moving fluids

Magnetic or gravitational fields

Plate tectonics or pressure gradients

Radioactive decays or spin configurations

The typical energies associated with each of these processes span a very wide range, which will certainly impact properties of life relying on them, such as their length scale and evolutionary timescale.

Based on our own example (and with the caveats appropriate for a sample of only one), we may also hypothesise that life requires a (molecular?) system compatible with Darwinian evolution and a (liquid?) environment allowing covalent bonds, although both of these are certainly questionable. One may also ask whether heavy elements are necessary for the emergence of life. This last question is the more directly relevant one in an astronomical context since, as we have already discussed, heavy elements such as carbon or oxygen are only available at the end of the life cycle of a first generation of stars.

Here as in other places, it is worth keeping in mind what is often called the mediocrity principle: whenever we thought that our place in the Universe was special, we found that it was not the case. We've known for a while that we are not at the center of the physical Universe, but many people (and even some observational astronomers) still believe that we are at the center of the biological Universe—a view sometimes know as carbon chauvinism.

In this context the mediocrity principle can be illustrated by the fact that the chemical composition of life on Earth is quite similar to the cosmological

Table 8.1 The chemical composition of the human body

Element	Mass (percent)	Atoms (percent)
Oxygen	65.0	24.0
Carbon	18.5	12.0
Hydrogen	9.5	62.0
Nitrogen	3.2	1.1
Calcium	1.5	0.2
Phosphorus	1.0	0.2
Potassium	0.4	<0.1
Sulphur	0.3	0.4
Sodium	0.2	0.4
Chlorine	0.2	0.2
Magnesium	0.2	0.1
Iodine	<0.1	<0.1
Iron	<0.1	<0.1
Zinc	<0.1	<0.1
Everything else	<1.0	<1.0

Percentages have been rounded off to one decimal place

abundances. The six most abundant elements are hydrogen, helium, oxygen, carbon, neon, and nitrogen. Of these, helium and neon are noble gases with very low chemical reactivity, while the other four make up almost 98% of the Earth's biomass and more than 96% of the human body, as show in Table 8.1.

One sometimes argues for a supposed superiority of carbon as a building block for life by pointing out that, of the 221 (at the time of writing) molecules that have been found in interstellar space, a large fraction are hydrocarbons. However, this is not a compelling argument, since this fraction merely reflects the comparatively high cosmological abundance of carbon. If one compares the fraction of molecules containing carbon and the fraction containing silicon (which is carbon's chemical neighbour in the periodic table), one finds that the two fractions are commensurate with the cosmological abundances of the two—as might have been expected.

Another frequent but spurious argument is that the Earth is optimised for life as we know it. This is not too different from the intelligent design arguments made in the context of biology. The truth is of course the opposite: life on Earth, though a process of evolution, has become adapted to its conditions. That said, the Earth does have some major advantages.

The first is the constant presence of liquid water. This requires a relatively large planet, and such a planet will usually possess a substantial atmosphere. However, such a planet cannot be arbitrarily large: a much bigger planet will have a larger surface gravity, implying that its crust will be thinner and the surface flatter. (This is why the largest mountains in the Solar System are on

Mars and Vesta.) In that case we will typically have a large and deep ocean with little continental surface. In the case of the Earth, about 71% of the surface is covered by water, and these percentages would be even larger for a larger planet.

The Earth's rotation period of 24 h is also moderate, at least if compared to those of Jupiter and Saturn (which rotate in about 10 h). This is useful because it minimises the possibility of extreme temperature differences between the day and night side of the planet (that is, illuminated or not by the Sun), or of violent weather patterns—hurricanes, for example, grow due to temperature differences between the equatorial and polar regions.

A third and crucial advantage of the Earth is its magnetic field (due to a region of liquid iron in its core), which deflects the solar wind and cosmic rays and preserves the atmosphere. Here Mars provides a good comparison point, because early on it also had a substantial atmosphere and liquid water on the surface. However, being about half the size of the Earth it cooled down faster. Eventually, the liquid core froze, which stopped the magnetic field, as well as element segregation, plate tectonics, and volcanism. Without the magnetic field, the solar wind gradually depleted the Martian atmosphere.

Mars and Venus are also good examples to illustrate the physical shortcomings of the concept of habitable zone. Firstly, basing habitability on the star's temperature and the planet's distance to the star has obvious problems if one simply remembers that the Moon and the Earth are (on average) at the same distance from the Sun. But Venus and Mars were both habitable in the past, and if one imagines the exercise of restarting the Solar System but swapping the positions of Venus and Mars, both would still be habitable today. On the one hand, if Venus had been further away from the Sun, it would not have developed a runaway greenhouse effect. On the other hand, if Mars had been closer to the Sun, it would still have a liquid iron core, due to the increased tidal effects from the Sun. These examples show that, if anything, when it comes to habitability, gravity is more important than atmosphere.

The presence of the Moon also plays a role, since it contributes to the stability of the Earth's axis of rotation. The inclination of this axis, which is currently about 23.5°, undergoes one of the so-called Milankovic cycles with a period of 41,000 years and an amplitude of only a few degrees. By comparison, Mars has two satellites (almost certainly captured asteroids) but these have much smaller masses and therefore a negligible dynamical effect, leading to an amplitude of tens of degrees in its analogous cycle.

Water itself is in many ways a remarkable substance, starting with the simple but important fact that the solid phase is less dense than the liquid phase—or in other words, that ice floats in liquid water. Without this property, ice would accumulate at the bottom of liquid water and therefore lakes and oceans would eventually become entirely frozen. Instead, a top layer of ice protects lower layers from additional cooling, allowing liquid water to remain available.

Other remarkable properties of water include having melting and boiling points much higher than those of analogous substances (only two others with comparable properties are known, ammonia and hydrogen fluoride), a large thermal capacity and conductivity (implying that it is an efficient environmental temperature stabiliser), an enthalpy higher than any other liquid (which makes it the ideal liquid for cooling by evaporation), a high solubility for polar molecules (and easy dissociation of their ions), and a high surface tension (which facilitates capillarity).

A final advantage of the Earth for carbon-based life forms is the presence of a stratospheric layer of ozone (O_3), which blocks ultraviolet radiation. Indeed, ultraviolet radiation is harmful for all carbon-based life. The evolution of life on Earth started in the oceans, and there is evidence that the migration to land occurred only after the formation of the ozone layer. Ultraviolet radiation is not a problem for ocean life since water is an efficient filter.

The search for life elsewhere in the Solar System is predominantly focused on Mars. For example, NASA has been regularly announcing discoveries of water on Mars since the Mariner 9 spacecraft in 1971. Nevertheless, other Solar System objects are potentially more interesting places to search. To find out whether an atmosphere is essential for the development of life, Europa (one of Jupiter's Galilean satellites, see the top panel of Fig. 8.12) and Enceladus (one of Saturn's satellites, discovered by William Herschel): neither of them has an atmosphere, but both have liquid water oceans underneath their frozen surfaces—and indeed there is more liquid water on Europa than on the Earth. To find out whether water is essential, a good place to search is Titan (Saturn's largest moon, discovered by Christiaan Huygens, see the bottom panel of Fig. 8.12), which has as a very substantial atmosphere: the surface pressure is almost 50% higher than that on the Earth.

But if there is life elsewhere, and indeed if it is common in the Universe, where is everybody? This is sometimes known as Fermi's paradox, after Enrico Fermi (1901–1954). This is an interesting question because an answer or its opposite could be correct. For example, we may be in the first planet in the Milky Way where intelligent life developed, but we might also be the last one. Maybe other more advanced civilizations know about us but do not want to interfere, either because we are too common or because we are too special (or fragile). It is also possible to develop consciousness without intelligence or intelligence without consciousness. And finally, life may exist on a much smaller or a much larger scale, and in such a case detecting it would be especially challenging for us.

Fig. 8.12   Top: Europa, imaged in 1996 by the Galileo spacecraft. Credit: NASA/JPL/DLR.
Bottom: Titan, imaged in 2012 by the Cassini spacecraft. Credit: NASA

To some extent, the best place to look for other life forms is on the
Earth itself. Such life forms are usually called shadow biosphere (composed
of descendants of a different origin), and the argument is that if life arose

independently twice on the same planet, it must be common throughout the Universe, whereas if it only arose once it is likely to be rarer. One can also debate the question of the number of times that intelligence has evolved on the Earth, bearing in mind dolphins and other species.

For several decades there has been a somewhat naive (though of course well-intentioned) effort to search for communications from alien civilisations. This was partially based on the premise that advanced civilizations would engage in radio communications, but it is clearly a parochial assumption: we ourselves did that extensively for a few decades, but we do it much less now. Curiously enough, in recent years there are arguments emerging that advanced civilisations should communicate using gravitational waves, and no doubt in a few years neutrinos or other means will be considered too.

If we wish to find civilizations that have the potential to communicate with us, we should try to estimate how often an industrial civilization develops, given that life has arisen and that some species are intelligent. It is not at all obvious that technology, science, or astronomy will necessarily develop in an otherwise intelligent society. Again, one just needs to imagine how our lifestyle would be, and the development of civilisation would have been, if the sky had been completely and permanently covered with thick clouds.

From a physics perspective, what might be the alternatives to life as we know it? The various possibilities that have been considered can be divided into four different classes, ordered from the least to the most different from ours:

Life analogous to life on Earth: this could be based on different nucleic acids and/or different chirality, or with the DNA replaced by other linear compounds.

Life with a different metabolism: for example, using sulphur instead of oxygen (thus necessarily having different amino acids and proteins), or metal-based for temperatures sufficiently high for metals to be liquefied (such as in exoplanets close to their parent stars, which is the case of most of those discovered so far).

Life with different chemistry: for example, using hydrocarbons as solvents, using reduction reactions—that is, hydrogenation—as an energy source (this may be relevant on Jupiter and Saturn); another possibility is life based on silicon or (for high temperatures) metal oxides, the latter again being relevant inside planets, and for planets very close to stars.

Life with different energy processes: for example, at the subatomic scale (relying on nuclear reactions in white dwarfs or neutron stars), on gases (atoms and molecules in interstellar space), on plasmas (using magnetic interactions in stellar atmospheres), or on solids (near absolute zero, using infrared radiation of molecular hydrogen spin isomers).

Finally, it is worth bearing in mind that the surface of a planet is (to some extent) the worst possible place for the long-term development of life. The gravitational field limits the size, neurological activity, and shape of living beings, and part of their resources are used for protection and movement. Moreover, these resources are finite, since the lifetime of the star which the planet is orbiting is also finite. The development of intelligence provides a selective advantage (we can better optimise the available resources), but how commonly is that stage reached?

Going back to the beginning of the section, today we don't know the answer to that question of whether or not we are alone in the Universe. But it is also important to bear in mind that the part of the Universe that we have explored in detail is tiny, even on the scale of our own galaxy. In the coming years this explored region will be significantly extended. While you wait for answers, some of the alternatives mentioned above have also been discussed in detail in the context of science fiction, where you can further explore them. A few nice examples are the following:

Isaac Asimov (1920–1992): Nightfall

Fred Hoyle (1915–2001): The Black Cloud

Arthur C. Clarke (1917–2008): Technical Error, Crusade

Robert Forward (1932–2002): Dragon's Egg, Starquake, Rocheworld (pentalogy), Camelot 30 K

Greg Egan (b. 1961): Diaspora, Orthogonal (trilogy)

A

Final Exam

Here I list some of the questions that have been previously included in written exams of courses I taught and which included some of the material in this book. The first part is a set of multiple-choice questions, each of which has only one correct answer. The second part is a list of possible essay topics, which in the context of the book you can take as themes for further discussion.

A.1 Multiple Choice Questions

Each of them has one and only one correct answer.

A.1. The scientific method is based on

(a) Consensus obtained through voting by the scientific community.
(b) The validation of hypotheses based on their underlying philosophy.
(c) The verification of hypotheses using experimentation and observation.
(d) The verification of hypotheses using argumentation.

A.2. Which of these characteristics is not essential in a scientific theory?

(a) Conceptual simplicity.
(b) Predictive power.

© Springer Nature Switzerland AG 2020
C. Martins, *The Universe Today*, Astronomers' Universe,
https://doi.org/10.1007/978-3-030-49632-6

(c) Falsifiability.
(d) Durability.

A.3. In terms of methodology, which of these is closest to astronomy?

(a) Astrology.
(b) Geology.
(c) Chemistry.
(d) Physics.

A.4. How many days would a year have if the Earth rotated with the same speed but in the opposite direction?

(a) One more.
(b) Two more.
(c) One less.
(d) The same.

A.5. For an observer on the Moon, how many times does the Earth rise and set during one thirty-day lunar month?

(a) None.
(b) One.
(c) Fifteen.
(d) Thirty.

A.6. Cervantes and Shakespeare died on 23 April 1616. Who died first?

(a) Shakespeare died 10 days after Cervantes.
(b) Shakespeare died 10 days before Cervantes.
(c) Shakespeare died 1 year after Cervantes.
(d) Shakespeare died 1 year before Cervantes.

A.7. Which of these calendars was luni-solar?

(a) The Egyptian calendar.
(b) The Islamic calendar.
(c) The Julian calendar.
(d) The Athenian calendar.

A.8. The seven-day week

(a) Is an Egyptian invention.
(b) Is a Babylonian invention.
(c) Is a Greek invention.
(d) Is a Roman invention.

A.9. Which of the following concepts was not introduced by the Pythagorean school?

(a) The sphericity of the Earth.
(b) Earth's rotation and translation movements.
(c) The separation between terrestrial and celestial regions.
(d) The 4 elements as a solution to the Problem of Change.

A.10. Models with epicycles were developed in order to

(a) Satisfy the Galilean concept of inertia.
(b) Find a simple alternative to heliocentrism.
(c) Explain why Mercury and Venus are always near the Sun.
(d) Reconcile observations with philosophical preferences for circular movements.

A.11. If the Earth was the center of the Universe, an astronomer on Mars would see

(a) Always the same face of the Earth.
(b) All planets, except the Earth, moving around Mars.
(c) The Earth moving around Mars, and the other planets moving around Earth.
(d) All planets moving around the Sun.

A.12. According to Aristotle, if you drop a lunar rock near Earth's surface, it will

(a) Float at a certain height above the ground.
(b) Rise towards the Moon.
(c) Fall towards the Earth.
(d) Have a behaviour that will depend on its weight.

A.13. An astronaut on the Moon releases a terrestrial rock. According to Aristotle, the rock

(a) Will fall towards the Earth.
(b) Will fall towards the Moon.
(c) Will stay somewhere between the Earth and the Moon.
(d) Will orbit the Moon at the height at which it was released.

A.14. In Western Europe around the year 1100

(a) The Earth's sphericity was considered doubtful.
(b) Aristotelian philosophy was dominant.
(c) Platonic philosophy had not yet been Christianised.
(d) A simplified version of Ptolemy's model was preferred.

A.15. In the model of Copernicus, an observer on Mars will see

(a) All planets moving around the Earth.
(b) Always the same face of the Earth.
(c) The Earth occasionally having a retrograde motion.
(d) That none of the planets has retrograde motion.

A.16. Which of Galileo's observations is necessarily incompatible with the model of Aristotle and Ptolemy?

(a) The phases of Venus.
(b) Sunspots.
(c) The satellites of Jupiter.
(d) The irregularities on the Moon's surface.

A.17. Galileo

(a) Is responsible for the current concept of inertial motion.
(b) Demonstrated the Earth's rotational movement.
(c) Defended the physical reality of the Copernican model.
(d) Demonstrated the impossibility of Tycho's model.

A.18. According to Galileo, an observer on Jupiter should see

(a) Both the Earth and Saturn having a complete cycle of phases.
(b) Saturn (but not the Earth) having a complete cycle of phases.
(c) The Earth (but not Saturn) having a complete cycle of phases.
(d) Neither the Earth nor Saturn having a complete cycle of phases.

A.19. An inertial observer

(a) Must have zero speed.
(b) Observes that bodies not subject to forces move with uniform speeds in rectilinear paths.
(c) Is at rest with respect to the center of the Earth.
(d) Has a uniform acceleration.

A.20. Shooting a bullet horizontally at the same time that another identical one is dropped from rest, which one reaches the ground first?

(a) The one that was shot.
(b) The one that was dropped.
(c) Both reach it at the same time
(d) There isn't enough information to answer.

A.21. Astronauts float inside the ISS because

(a) There is no gravity in space.
(b) They are in a very weak gravitational field.
(c) They are constantly falling.
(d) Their inertial mass is zero inside the ISS.

A.22. A consequence of Galileo's principle of relativity is that

(a) Velocities are absolute.
(b) Accelerations are relative.
(c) The speed of sound is constant.
(d) Velocities are relative.

A.23. In an inertial frame of reference one observes that

(a) Bodies do not move at all.
(b) Bodies are affected by gravity independently of their mass.
(c) Bodies not affected by forces move at constant speed in a straight line.
(d) Bodies not affected by forces move at constant acceleration in a straight line.

A.24. A consequence of Newton's assumption that space and time are absolute is that

(a) All speeds must be relative.
(b) Some lengths can be relative.
(c) Simultaneity must relative.
(d) The speed of light must be absolute.

A.25. Two inertial observers measure the speed of the same beam of light and obtain the same finite value. This would surprise Newton because

(a) According to him the speed of anything should depend on the frame of reference.
(b) He expected the speed of light to be infinite in inertial frames.
(c) In inertial reference frames he believed light to be stationary.
(d) He believed light should have an infinite speed in all frames.

A.26. For which planet are the effects of relativity more important?

(a) Mercury.
(b) Venus.
(c) Uranus.
(d) Neptune.

A.27. Geosynchronous satellites

(a) Must be above the Equator.
(b) Are used in the GPS system.
(c) Can never observe the North and South Poles.
(d) None of the above.

A.28. A spy satellite wanting to photograph your house must be orbiting at a distance

(a) Very close to the Earth.
(b) A few times the Earth's radius.
(c) Comparable to that of the Moon.
(d) Of the L1 or L2 Lagrange points.

A.29. Special relativity (SR) and general relativity (GR) corrections must be taken into account for the GPS system to work. If these satellites were moved to LEO,

(a) Both corrections would decrease.
(b) The SR correction would decrease and the GR one would increase.
(c) The SR correction would increase and the GR one would decrease.
(d) Both corrections would increase.

A.30. A property of Maxwell's equations is that

(a) Some solutions represent waves whose speed is independent of the observer.
(b) Some solutions represent waves whose speed depends on the observer.
(c) They imply that all charged objects move at the speed of light.
(d) They are perfectly compatible with Newtonian physics.

A.31. The Michelson–Morley experiments were an attempt to

(a) Measure the speed of light.
(b) Test the Lorentz–Fitzgerald transformations.
(c) Measure the Earth's speed relative to the Sun.
(d) Detect the Earth's movement relative to the aether.

A.32. Suppose that you verify experimentally that a certain particle moves at the speed of light. This implies that

(a) Its energy is zero.
(b) Its rest mass is zero.
(c) Its energy is infinite.
(d) Special relativity must be wrong.

A.33. Special relativity

(a) Is based on the universality of free fall.
(b) Is based on the equivalence principle.
(c) Implies the equivalence principle.
(d) Implies the non-existence of the aether.

A.34. Relative to an object at rest, a moving object is

(a) Shorter and younger.
(b) Longer and older.
(c) Shorter and older.
(d) Longer and younger.

A.35. Electrons can't move at the speed of light because

(a) That would violate the equivalence principle.
(b) Their size would become infinite.
(c) That would violate the principle of relativity.
(d) That would require an infinite energy.

A.36. The equivalence principle implies that

(a) The speed of light is absolute.
(b) The inertial and gravitational mass are identical.
(c) All gravitational forces vanish in inertial frames of reference.
(d) The gravitational acceleration is proportional to the product of the two masses.

A.37. If an astronaut in a windowless rocket feels pressed against the back of her seat, she won't be able to tell

(a) Whether the rocket is accelerating or moving at constant speed.
(b) Whether she is accelerating or resting on the surface of some planet.
(c) Whether the rocket it moving at constant speed in space or at rest in space.
(d) Whether she is at rest in space or freely falling towards some planet.

A.38. A green animal lives on an extremely dense planet. When observed from space,

(a) It will look green.
(b) It will look white.
(c) It will look blue.
(d) It will look red.

A.39. The Hertzsprung–Russell diagram is a way of classifying stars based on their

(a) Mass and luminosity.
(b) Temperature and radius.
(c) Radius and mass.
(d) Luminosity and color.

A.40. A pulsar is

(a) A rotating neutron star which emits a strong X-ray beam.
(b) A main sequence star whose size increases and decreases periodically.
(c) A white dwarf in a binary system, whose luminosity changes regularly.
(d) A rotating black hole which emits a strong microwave beam.

A.41. In white dwarfs

(a) Thermonuclear reactions rapidly transform all helium into carbon.
(b) There is no mechanism capable of balancing gravity.
(c) The gravitational pull is balanced by the electron degeneracy pressure.
(d) The centrifugal force leads to strong magnetic fields.

A.42. A star with a total mass of 7 solar masses

(a) Will end its life as a white dwarf.
(b) Produces almost all its energy through pp chains.
(c) Has a radiative core and a convective exterior.
(d) Will produce some elements heavier than iron.

A.43. To turn the Sun into a black hole we would need to compress its mass into a size of

(a) A few millimetres.
(b) A few metres.
(c) A few kilometres.
(d) The Earth.

A.44. The cosmic redshift of the galaxies

(a) Is due to the motion of galaxies in space.
(b) Is due to the Doppler effect.
(c) Is due to the expansion of space.
(d) Has been predicted but not yet observed.

A.45. Which of these elements was not formed at the beginning of the Universe?

(a) Hydrogen.
(b) Helium.
(c) Lithium.
(d) Carbon.

A.46. The current age of the Universe is about

(a) Probably infinite.
(b) 400 million years.
(c) 14 billion years.
(d) 140 billion years.

A.47. According to recent observational data, the rate of expansion of the Universe

(a) Is constant.
(b) Is accelerating.
(c) Is decelerating.
(d) Is very small.

A.48. Most of the matter in the Universe is

(a) Hydrogen.
(b) Neutrinos.
(c) Black Holes.
(d) Dark matter.

A.49. The discovery of dark energy necessarily implies that

(a) There must be a significant amount of antimatter.
(b) The expansion of the Universe is temporarily accelerating.
(c) General relativity is incorrect.
(d) The equivalence principle is incorrect.

A.50. Which of these provides supporting evidence for the Big Bang model?

(a) The cosmic microwave background.
(b) The absence of antimatter.
(c) Galactic rotation curves.
(d) The existence of dark energy.

A.51. The main cosmological implication of Type Ia supernova observations is that

(a) There must be a significant (but still unknown) amount of antimatter.
(b) The Universe is currently dominated by a fluid with negative pressure.
(c) General relativity must be incorrect.
(d) The Einstein equivalence principle must be violated.

A.52. Which of these is not a direct test of the Einstein equivalence principle?

(a) Measurements of gravitational redshift.
(b) Tests of the stability of fundamental couplings.
(c) Measurements of the precession of Mercury's perihelion.
(d) Tests of the universality of free fall.

A.53. Which of the following can't (as far as we know) be explained by extra dimensions?

(a) The fact that gravity is the weakest of the fundamental forces.
(b) The unification of the four fundamental forces.
(c) The small value of the cosmological constant.
(d) The spacetime variation of fundamental constants.

A.54. The formation of cosmological topological defects is a consequence of the fact that in our Universe

(a) There are 3 spatial dimensions.
(b) There is an odd number of spatial dimensions.
(c) The speed of light is finite.
(d) The age of the Universe is finite.

A.55. A conclusive detection of spacetime variations of the fine-structure constant would necessarily imply that

(a) The Universe is not accelerating.
(b) Gravity is not purely geometry.
(c) Dark energy must be a cosmological constant.
(d) String theory must be correct.

A.56. The redshift drift of objects following the cosmological expansion

(a) Is always positive.
(b) Is positive for decelerating universes and negative for accelerating universes.
(c) Is negative for decelerating universes and positive for accelerating universes.
(d) Is always negative.

A.57. The violation of Bell's inequality

(a) Rules out theories with hidden variables.
(b) Proves that quantum mechanics is incomplete.
(c) Implies that the Universe must have extra dimensions.
(d) Was predicted by Einstein but hasn't yet been measured.

A.58. The most common element in the human body is

(a) Carbon (by number).
(b) Carbon (by mass).
(c) Hydrogen (by mass).
(d) Oxygen (by mass).

A.59. The fact that hydrocarbons are very common among molecules found in the interstellar medium

(a) Reflects the cosmological abundance of carbon.
(b) Is abnormal given the abundances of carbon and hydrogen.
(c) Is due to the fact that we preferentially search for them.
(d) Shows that carbon is ideal for life.

A.60. In the Solar System, water in liquid form

(a) Only exists on the Earth.
(b) Exists in many other bodies, but only in small amounts.
(c) Exists only on the Earth, Europa, and Enceladus.
(d) Mainly exists underneath layers of ice.

A.2   Essay Topics

Write a short essay (of about one page) on each of the following topics.

A.61. What are the differences between scientific knowledge and opinion? Discuss an example from the book, or alternatively from current news, to illustrate your answer.

A.62. Imagine yourself living in a primitive society and without access to any technological means. How might you try to convince them that the Universe is predictable and governed by a set of laws? What objections or counter-arguments might you find?

A.63. Imagine yourself living in Classical Greece. How would you try to convince them that the same laws of physics apply on Earth and in the rest of the Universe? How would you try to convince them that the planetary motions need not be described by circles?

A.64. Discuss the role of astronomy and physics in the development of our society and culture, and in particular what their role is (or should be) in our modern society. Illustrate your opinion with some specific examples.

A.65. Compare possible definitions of simplicity and/or beauty for an astronomer (or a physicist) and a musician. In the former case, discuss how these may be influenced or changed by experimental results.

A.66. Discuss some similarities and differences between astronomy and other sciences, such as physics, chemistry, geology, or biology. Illustrate your ideas with one or more examples in the book.

A.67. Discuss how astronomy, and society as a whole, might have developed if the Earth had no Moon.

A.68. Discuss how astronomy, and society as a whole, would have developed if the Sun was part of a binary system. You may assume, for example, that Jupiter is replaced by a red dwarf star.

A.69. Discuss how astronomy and physics would have evolved if Galileo had discovered the planet Neptune while studying Jupiter's satellites.

A.70. Discuss what might be our current vision of the Universe if the sky was always covered with thick clouds. Speculate about possible implications for other sciences and for society as a whole.

A.71. Discuss what would happen to the Earth if the Sun suddenly became a black hole. You should consider both short-term and long-term effects.

A.72. Discuss what might be our current vision of the Universe if astronomy relied exclusively on naked eye observations. Speculate about possible implications for other sciences and for society as a whole.

A.73. Discuss how life on the Earth might be different if it was tidally locked with the Sun (just as the Moon is tidally locked with the Earth). Speculate on possible impacts on the development of astronomy and physics.

A.74. Discuss the history of the discovery of the planet Neptune as an example of the validity of the scientific method in astronomy. You may discuss similarities and differences with astrology, or with the non-discovery of the planet Vulcan.

A.75. Does the development of scientific ideas and knowledge necessarily require experiments? Whether you want to argue for yes or no, explain why and use one or more examples from the book to defend your position.

A.76. Discuss the way in which the theory of relativity has changed our view of the Universe, and compare that change with the one that occurred when Aristotelian physics was replaced by Newtonian physics.

A.77. Describe the relation between physical reality and the mathematical models with which we describe it. In particular, comment on how this relation may differ in the contexts of the pure and applied physical sciences (e.g., theoretical physics and engineering).

A.78. Should ideas such as string theory or the multiverse, for which there is (at least currently) absolutely no evidence, be considered scientific? Whether you want to argue for yes or no, explain why and use one or more examples from the book to defend your position.

A.79. The anthropic principle is the philosophical point that any observation of the Universe that we make must be compatible with our existence. Is this useful as a physics concept? Whether you want to argue for yes or no, explain why and use one or more examples from the book to defend your position.

A.80. It has been said that the multiverse is for modern physics (or cosmology) what intelligent design is for modern biology, because both have zero evidence supporting them. Whether or not you agree, discuss the reasons for such an analogy.

The answers to the multiple choice questions can be found online at https://www. dropbox.com/s/ddfjlcc76kgnvuq/TheUniverseToday.pdf?dl=0.

B

Further Reading

The following is an alphabetic list of useful bibliographic references and/or suggestions for further reading. Note that there is a range of levels of difficulty, from classic and/or popular science books (e.g., Sagan's Cosmos) to modern technical volumes.

Aaboe, A.: Episodes From the Early History of Astronomy (Springer, 2001)

Bais, S.: Very Special Relativity: An Illustrated Guide (Harvard University Press, 2007)

Barrow, J.D.: The Constants of Nature: The Numbers That Encode the Deepest Secrets of the Universe (Vintage, 2004)

Barrow, J.D., Tipler, F.J.: The Anthropic Cosmological Principle (Oxford University Press, 1988)

Bartusiak, M.: Einstein's Unfinished Symphony: Listening to the Sounds of Space-Time (Joseph Henry Press, 2000)

Bartusiak, M.: Archives of the Universe: 100 Discoveries that Transformed our Understanding of the Cosmos (Vintage Books, 2006)

Bernal, J.D.: The extension of man: A history of physics before 1900 (Weidenfeld and Nicolson, 1972)

Bricmont, J.: Quantum Sense and Nonsense (Springer, 2017)

Burke, J.: The Day the Universe Changed (BBC Books. 1985)

Christianson, G.E.: Isaac Newton (Oxford University Press, 2005)

Crelinsten, J.: Einstein's Jury: The Race to Test Relativity (Princeton University Press, 2006)

Cushing, J.T.: Philosophical Concepts in Physics: The Historical Relation Between Philosophy and Scientific Theories (Cambridge University Press, 1998)

© Springer Nature Switzerland AG 2020
C. Martins, *The Universe Today*, Astronomers' Universe,
https://doi.org/10.1007/978-3-030-49632-6

Drake, S.: Galileo: A Very Short Introduction (Oxford University Press, 2001)

Dreyer, J.L.: A History of Astronomy from Thales to Kepler (Dover Publications, 1953)

Feynman, R.P.: The Character of Physical Law (The MIT Press, 2017)

Finocchiaro, M.: The Galileo Affair: A Documentary History (University of California Press, 1989).

Gingerich, O.: Nicolaus Copernicus: Making Earth a Planet (Oxford University Press, 2005)

Gottlieb, A.: The Dream of Reason: A History of Western Philosophy from the Greeks to the Renaissance (W.W. Norton, 2016)

Gribbin, J.: Science: A History (Penguin Books 2010)

Grosser, M.: The Discovery of Neptune (Dover Publications, 1979)

Hannam, J.: God's Philosophers: How the Medieval World Laid the Foundations of Modern Science (Icon Books, 2010)

Heilbron, J.L.: Galileo (Oxford University Press, 2015)

Heilbron, J.L.: Physics: A Short History from Quintessence to Quarks (Oxford University Press, 2012)

Hoskin, M.: Discoverers of the Universe: William and Caroline Herschel (Princeton University Press, 2011)

Jones, A.: A Portable Cosmos: Revealing the Antikythera Mechanism, Scientific Wonder of the Ancient World (Oxford University Press, 2017)

Kanas, N.: Star Maps: History, Artistry, and Cartography (Springer, 2012)

Kennefick, D.: No Shadow of a Doubt: The 1919 Eclipse That Confirmed Einstein's Theory of Relativity (Princeton University Press, 2019)

Koestler, A.: The Sleepwalkers: A History of Man's Changing Vision of the Universe (The Danube edition, Hutchison of London, 1979)

Kragh, H.: Higher Speculations: Grand Theories and Failed Revolutions in Physics and Cosmology (Oxford University Press, 2015)

Lequeux, J: Le Verrier: Magnificent and Detestable Astronomer (Springer, 2013)

Levenson, T.: The Hunt for Vulcan: …And How Albert Einstein Destroyed a Planet, Discovered Relativity, and Deciphered the Universe (Random House, 2016)

Liddle, A.: An Introduction to Modern Cosmology (Wiley, 2003)

Lindberg, D.C.: The Beginnings of Western Science: The European Scientific Tradition in Philosophical, Religious and Institutional Context, 600 BC to AD 1450 (University of Chicago Press, 1992)

Linton, C.M.: From Eudoxus to Einstein: A History of Mathematical Astronomy (Cambridge University Press, 2004)

MacLachlan, J.: Galileo Galilei: First Physicist (Oxford University Press, 1997)

Marchant, J.: Decoding the Heavens: Solving the Mystery of the World's First Computer (Windmill, 2009)

Martins, C.J.A.P.: Cosmology with Varying Constants, in: Advances in astronomy, Thompson, J.M.T. (ed.), Royal Society Series on Advances in Science, Vol. 1, pp. 41–58 (Imperial College Press, 2005)

Martins, C.J.A.P.: Defect Evolution in Cosmology and Condensed Matter: Quantitative Analysis with the Velocity-Dependent One-Scale Model (Springer, 2016)

Martins, C.J.A.P.: The status of varying constants: a review of the physics, searches and implications, Reports on Progress in Physics, Volume 80, article id. 126902 (2017)

Mermin, N.D.: It's About Time: Understanding Einstein's Relativity (Princeton University Press, 2009)

Muller, R.A.: Physics and Technology for Future Presidents: An Introduction to the Essential Physics Every World Leader Needs to Know (Princeton University Press, 2010)

Naess, A.: Galileo Galilei: When the World Stood Still (Springer, 2005)

North, J.D.: The Fontana History of Astronomy and Cosmology (Fontana Press, 1994)

Pagels, H.R.: Perfect Symmetry: The Search for the Beginning of Time (Simon & Schuster, 2009)

Pagels, H.R.: The Cosmic Code: Quantum Physics as the Language of Nature (Dover Publications, 2012)

Panek, R.: The 4 Percent Universe: Dark Matter, Dark Energy, and the Race to Discover the Rest of Reality (Mariner Books, 2011)

Paulos, J.A.: Innumeracy: Mathematical Illiteracy and Its Consequences (Hill and Wang, 1989)

Pecker, J.-C.: Understanding the Heavens: Thirty Centuries of Astronomical Ideas from Ancient Thinking to Modern Cosmology (Springer, 2001)

Pedersen, O.: Early Physics and Astronomy: A Historical Introduction (Cambridge University Press, 1993)

Richards, E.G.: Mapping Time: The Calendar and Its History (Oxford University Press, 1998)

Russell, B.: ABC of Relativity (Routledge, 2009)

Russell, J.B.: Inventing the Flat Earth: Columbus and Modern Historians (Praeger, 1991)

Russo, L.: The Forgotten Revolution: How Science Was Born in 300 BC and Why it Had to Be Reborn (Springer, 2004)

Sagan, C.: Cosmos (Ballantine Books, 1985)

Sagan, C.: The Demon-Haunted World: Science as a Candle in the Dark (Ballantine Books, 1997)

Sambursky, S.: The Physical World of the Greeks (Routledge and Kegan Paul, 1956)

Simonyi, K.: A cultural history of physics (CRC Press, 2012)

Thoren, V.E.: The Lord of Uraniborg: A Biography of Tycho Brahe (Cambridge University Press, 1991)

Vilenkin, A., Perlov, D.: Cosmology for the Curious (Springer, 2017)

Voelkel, J.R.: Johannes Kepler and the New Astronomy (Oxford University Press, 2001).

Webb, J.K. et al.: Indications of a spatial variation of the fine structure constant, Physical Review Letters, Volume 107, article id. 191101 (2011).

Weinberg, S.: The First Three Minutes (Basic Books, 1977)

Westman, R.S.: The Copernican Question: Prognostication, Skepticism, and Celestial Order (University of California Press, 2011)

Whitrow, G.J.: Time in History: Views of Time from Prehistory to the Present Day (Oxford University Press, 1989)

Wilczynska, M.R. et al.: Four direct measurements of the fine-structure constant 13 billion years ago, Science Advances, Volume 6, eaay9672 (2020)

Will, C.M.: Was Einstein Right? Putting General Relativity to the Test (Basic Books, 1993)

Wootton, D: Galileo: Watcher of the Skies (Yale University Press, 2013)

Wudka, J.: Space-time, Relativity and Cosmology (Cambridge University Press, 2006)

Wynn-Williams, G.: Surveying the Skies: How Astronomers Map the Universe (Springer, 2015)

Index

© Springer Nature Switzerland AG 2020
C. Martins, *The Universe Today*, Astronomers' Universe,
https://doi.org/10.1007/978-3-030-49632-6